GLIMPSE OF THE
OCEANOGRAPHY

GLIMPSE OF THE OCEANOGRAPHY

Shuchi Bhatt & Devanshi Joshi

www.whitefalconpublishing.com

Glimpse of the Oceanography
Shuchi Bhatt & Devanshi Joshi

www.whitefalconpublishing.com

All rights reserved
First Edition, 2022
© Shuchi Bhatt & Devanshi Joshi, 2022
Cover design by White Falcon Publishing, 2022
Cover image source freepik.com

No part of this publication may be reproduced, or stored in a retrieval system, or transmitted in any form by means of electronic, mechanical, photocopying or otherwise, without prior written permission from the author.

The contents of this book have been certified and timestamped on the Gnosis blockchain as a permanent proof of existence. Scan the QR code or visit the URL given on the back cover to verify the blockchain certification for this book.

The views expressed in this work are solely those of the author and do not reflect the views of the publisher, and the publisher hereby disclaims any responsibility for them.

Requests for permission should be addressed to
shuchimarine@gmail.com

ISBN - 978-1-63640-729-6

PREFACE

Oceans have always fascinated humankind. The present book entitled "**Glimpse of the Oceanography**" has covered various aspects of oceanography. This book has been written to endow the details about diverse features of oceanography and thus would help the students, teachers, and research scholars. To study oceanography one has to clear his/her basic concept about it. We sincerely believe that this book will help the students to understand the subject thoroughly.

In the first chapter of "Introduction", we have given brief information on different aspects of the ocean, like information on oceanographic history, Physical Oceanography, Chemical Oceanography, Biological Oceanography, Zonation, etc. Pacific Ocean, Atlantic Ocean, Indian Ocean, Southern Ocean, and Arctic ocean have been given in the second chapter of this book. The Third chapter "Habitat" has been elaborately described including Rocky Shore, Beaches, Mud Flats, Intertidal Zone, Estuaries, Coral reef, Mangrove, Seagrass, etc. The fourth chapter "Marine Life" described marine Invertebrate, Vertebrate and marine vegetation. As we have already mentioned in an earlier chapter that coral reef is an important factor to be noticed, we have included a separate chapter (chapter 5).

We have endeavored to make this book scientifically accurate; however, We bear the entire responsibility for any mistakes.

- Shuchi Bhatt
- Devanshi Joshi

MESSAGE

The Ocean plays a significant role in our lives and has done so for a long time. The survival of marine and coastal ecosystems and biodiversity is important to the nutritional, spiritual, societal, and religious well-being of many coastal communities. But even for the many millions of humans who may not think that they have any well-built dependence on the ocean, marine ecosystems provide all kinds of reimbursement. The survival of marine and coastal ecosystems and biodiversity is essential to the nutritional, spiritual, societal, and religious welfare of numerous coastal communities.

The present book "**Glimpse of the Oceanography**" includes a holistic approach to study various detailed studies such as a Basic introduction, Oceans of the world, Marine habitat, Marine life, and Coral reef. This book provides an important source for Students, Researchers, and the people of India to understand the incredible and indispensable role oceans play in our lives. Importantly the book highlights why we as a nation should preserveand protect our oceans!

Dr. Shaili Johri
Stanford University.

The book "**Glimpse of the Oceanography**" concisely presents the essentials of ocean science covering a wide range of topics dealing with oceans, ocean processes and ocean life. It discusses the realms of physical, chemical and biological oceanography in the context of both biological and ecological processes that play a crucial role in sustaining ocean productivity and ocean health. The marine habitats exhibit great diversity and host enormous biological diversity with potential economic, ecological and social interests. The book describes all major elements of oceanography along with relevant illustrations representing oceanography including the oceans of the world, marine life, marine habitat, coral reef, etc. It is a very good resource book and provides fundamental aspects of oceanography, particularly to all those dealing with oceans and ocean resource management.

CRK Reddy, Ph.D
Former Chief Scientist, CSIR-CSMCRI and
Adjunct Professor, ICT, Mumbai

Chapter 1
Introduction

The oceans are well recognized through their enormous spread and immense potential. The vastness and dynamic environment, exhibited by the ocean, have fascinated man for centuries. However, the curiosity and exploration of man remained limited up to the shores until the seamanship and navigation technologies developed an association of mankind with the oceans evolved gradually and, today, it has harmoniously intervened. It is often the blend of beauty, necessity, mystery, and variety of life in the sea that attracts us to excogitate marine realms more and more. The continuous rise and fall of a huge water mass, its synchrony with lunar cycles, its climatic and aesthetic influence on human life, and often, the gracefully leaping dolphins may have remained the earlier fascinating calls to the oceanographic explorations. Finally, the curiosity reached its ridge point and got supported by the technology which opened doors to a wonderful world of marine expeditions. Hence, the systematic collection of such expeditions and investigations gave rise to the novel fields of science referred to as Oceanography. It includes all aspects of the ocean i.e., physical, chemical, biological, geological as well as the evolution of oceans from Tethys sea to the present-day oceans; however present discussion will focus on the combination of the above-mentioned aspects i.e., marine ecology with special reference to biodiversity.

Some enthusiastic explorers carried out marine expeditions in past and provided us with the oceanographic facts, we have today. As far as oceanography is concerned, Matthew Maury(1806-1873) is revered undoubtedly for their exceptional contribution in this field, who is considered as 'Father of Modern Oceanography and Naval Meteorology'. His discoveries on currents and wind mapping could considerably reduce the long voyage time for traders and sailors. Some other oceanographer includes James Cook, Vagn Ekman, Sir Wyville Thomson, Sylvia Alice Earle, Henry B. Bigelow, Eugenie Clark.

A British explorer, James Cook (1728-1779) for the first time mapped the Pacific ocean accurately as well as could locate the Great Barrier Reef which holds the biodiversity habitat and the largest natural edifice amidst the ocean. He also confuted the existence of the southern continent of Australia. One of the characteristic features of oceans is currents, which were explained by Vagn Ekman (1874-1954) along with the impact of temperature and pressure on it to formulate the ocean compressibility. These ocean currents were further studied by Jacques Piccard (1922-2008), and he for the first time studied the deepest trench of the Pacific ocean. There is a long list; however, this segment aims to point out key contributors towards the marine biodiversity explorations. One of such is Sir Wyville Thomson (1830-1882) who refused the myth of "No life beneath Sea" and led the first exclusively scientific expedition to the deep sea. His journey with HMS *Challenger* contributed to the discoveries of many marine species along with physical and chemical data collection of that region. Sylvia Alice Earle (1935-Present) is a unique name in the science of oceanography, she is the pioneer of SCUBA diving, holds a world record of women's deepest untethered dive. She was the first woman to hold the position of a chief scientist at National Oceanic and Atmospheric Administration and she worked upon marine algae as well as generated awareness regarding overfishing and pollution. Charles William Beebe (1877-1962) was a marine biologist to explore deep-sea life. Henry Bigelow (1879-1967), was a marine biologist and gave a significant contribution to the study of cnidarians and elasmobranch, he also decipher complex marine ecology. Eugenie Clark (1922-2015) was more commonly known as a Shark lady due to her remarkable contribution to studying shark behaviour. There is a long list of contributors in this field; we pay our sincere gratitude to all who made it possible for us to understand the fascinating marine biology. Marine life offers countless resources to mankind, i.e., food, medicines, and raw materials.

The water is chief competent of all solvents which carry necessary solutes like essential gases, oxygen, and carbon dioxide. On our planet, our life commences with water; it is a precious source for living organisms. The sea level rise and fall vary at different places as well as time. Each trice of waves is hitting the coastline reflecting the dynamism of the ocean ecosystem over the terrestrial one. Marine organisms range from microscopic to gigantic from marine bacteria to the largest animal found in the living world i.e., Blue whale. Moreover, the number of species is increasing with discovery.

1.1 *Physical oceanography*

It is the study of physical conditions and processes of the ocean. It especially studies the motions and physical properties of ocean waters. Physical oceanography is subdivided into *descriptive* and *dynamic* physical oceanography. Physical oceanography focuses on recounting and explaining the evolving patterns of ocean circulation and fluid motion, along with the distribution of its properties such as temperature, salinity, and the concentration of dissolved chemical elements and gases, etc. Marine organisms cannot change their external environments, they have to either adjust with the surrounding environments or move elsewhere. Temperature, pH, calcium carbonate ($CaCO_3$), etc. are among the most important environmental factors controlling the distribution, physiological performance, morphology, and behavior of marine biodiversity.

In the marine ecosystem physical as well as chemical components are vital for living organisms. Physical parameters include density, temperature, salinity, water transparency, odor, water circulation, etc; while chemical parameters include dissolved gases, pH, nutrients, etc. The density of surface water is varied from 1020 kg/m^3 to 1029 kg/m^3, temperature and salinity affect density. The temperature of the water is different from sea to sea. Temperature is ranging -2°C to 30°C, at polar seas -2°C while at the Persian Gulf, it reaches up to 30°C.

The ocean ecosystem is wealthy for life victuals and scientific research. The hasty industrialization beside the coastal area has brought extensive declination of water quality particularly of brackish water and the estuaries, producing increased pressures foremost to environmental stress or even affecting public health.

- *Ocean currents and Circulation*

The ocean currents and circulation are the incessant, predictable, directional movement of seawater driven by gravity, wind, and density. Horizontal movements are referred to as currents, while vertical changes are called upwellings or downwellings. Mainly ocean circulations are divided into two types: Wind-driven circulation and Thermohaline circulation.

1.2 Chemical oceanography

It is the study of ocean chemistry based on the circulation and dynamics of fundamentals, isotopes, molecules, etc.

It includes organic as well as inorganic chemistry. The chemicals found in seawater are natural as well as of anthropogenic origin. Several phenomena contributed to the ocean chemistry such as the environmental impact of mass oil spills, carbon cycle, climate and paleoclimate, redox process, hydrothermal process, ocean mixing, etc. Salinity refers to the total amount of dissolved salts in water. salinity contains a variety of salts i.e., chloride(Cl^-) sodium(Na^+) sulfate (So_4^{-2}) magnesium(Mg^{+2}) etc. The salinity affects marine organisms, most marine biotas are sensitive to salinity change. The negligible changes in salinity might be harmful to marine life. Mostly the oceans have a 35% (ppt), which may show minor changes. Apart from this, nutrients such as nitrogen, phosphorus, potassium, sulfur, magnesium, etc. constitute chemical elements, essential for the growth and survival of marine biota.

There are solids as well as gasses that are dissolved in seawater. For living organisms oxygen, nitrogen, and carbon dioxide are crucial gases. The significant amounts of gases are different in the atmosphere and oceans. In the marine ecosystem, cold water can dissolve more amounts of gases than warm water. The primary producers intake carbon dioxide through photosynthesis reactions and produce oxygen for other organisms. Dissolved oxygen is necessary to various living organisms. Its amount is different from individual life forms. The reduced level of oxygen may lead to impairment in marine life. The average value of oxygen, 5 mg/l or above is considered as ultimately dissolved oxygen ranges. This supports a different biological process. The amount of dissolved oxygen is highest near the polar sea and lowest near the equator. If the temperature increases, the dissolved oxygen is decreased, while the temperature decreases the dissolved oxygen is increased. The pH of water is one of the most important components. The pH of seawater ranges 7.8-8.3. Apart from this range, acidic or alkaline kinds of sea water are harmful to marine organisms. The increase of acidity leads to the damage ocean ecosystem, coral might die or bleach. Nutrients are important for living organisms. Sunlight enters through the water column. The amount of sunlight depends on the transparency of water. A level of transparency is essential for photosynthesis reactions. Various water parameters like Dissolved Oxygen (DO), Biological Oxygen Demand (BOD), pH, Total Dissolved Solids (TDS), Total Suspended Solids (TSS), temperature, color, and salinity are the indicators of seawater quality.

1.3 Biological oceanography

It is the study of interaction with living organisms of the ocean. It is also referred to as marine ecology. Ocean renders 90% of the habitable space on earth to numerous wonderful creatures. Biological oceanography includes various studies such as microscopic plants, algae, animals, etc; biological oceanographers study all varieties of oceanic progression that involve living organisms. The study of marine biology by humans began as early as fourth century BC when Aristotle illustrated 180 species of marine animals. The geographical information of oceans got improved after several great sea expeditions conducted by the people from the 15th to 16th centuries. The biological exchanges among water, air, and life are studied under the branch of biological oceanography. These include progressions that occur at molecular scales, such as photosynthesis, respiration, and cycling of essential nutrients, too.

Marine life has 300 times more space available than other aquatic or terrestrial life living on the ground. Marine organisms are classified according to the marine environments they occupy. Oceanic species and neritic species are depending upon whether the organisms are found offshore or coastal waters respectively. It is believed that ancient organisms originated from marine water. In marine water, several micros to macro living organisms are found. In the ocean phytoplankton, zooplankton, seaweed, kelps, invertebrate, and vertebrates are found. They inhabit the intertidal zone to the twilight zone. The typical marine environment and the biotic community inhabiting the area collectively form various ecosystems in the marine realm. The biotic components include producers, consumers, predators, parasites, competitors, and mates. The abiotic components are temperature, the concentration of nutrients, sunlight, turbulence, salinity, and density.

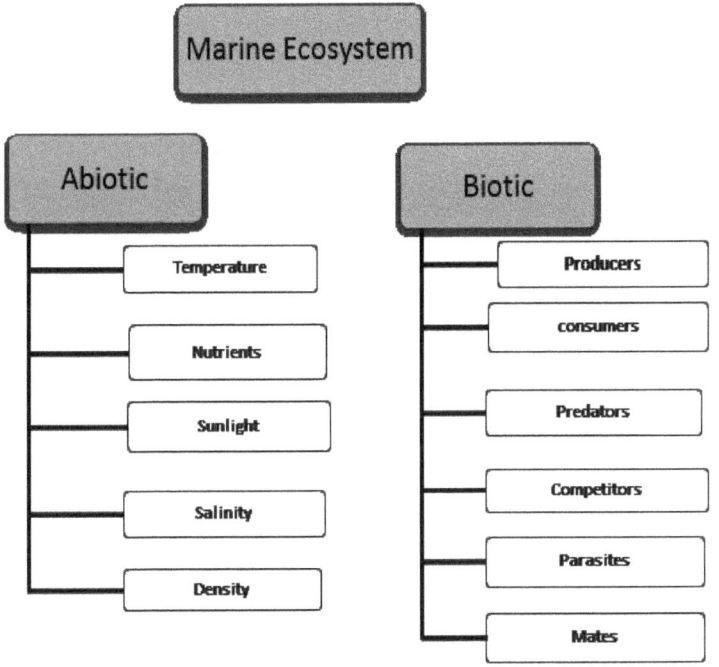

Image 1-1: Marine biotic and Abiotic factors

The marine ecosystems elucidate vital factors in the form of a spacious range of habitats. The foremost habitats such as mangroves, reefs, seaweeds, estuaries, etc. serve as an important survival ground for corals, fishes, etc. Humans too began colonizing in the ocean vicinity, since ancient times. The ecological study is elemental to the understanding of biology, as organisms cannot survive as isolated units. Important marine ecological studies are to understand marine ecosystems as working processes. There exists a wealth of biodiversity in marine environments as importance of millions of years of evolutionary history.

In the shallow water body, the abundance of marine life is vastly found due to sufficient light penetration. The commonly shallow water body has a light sufficiency it reaches well up to the bottom level. The total numbers of marine living species are less than that of the terrestrial ecosystem. Because the marine diversity has not been fully recorded, it is not easy to explore the ocean and collect samples. The Marine ecosystem is very deep in which existence extends to the entire depths. The water mass is a continuous movement in both vertical and horizontal dimensions. This ecosystem is characterized by various zones like a pelagic ocean, neritic ocean, and continental shelf. Variations of geological, chemical, and physical characters produce complex habitat structures. Intertidal zones are transition lines between land and ocean. The intertidal zone was exposed during low tide moreover submerged during the high tide. Organisms that live within the tidal pools are adapted to tremendous changes in salinity. They can sustain with falling and rising salinity levels. During the monsoon, freshwater mixes with saline water and reduces the salinity of the respective marine ecosystem.

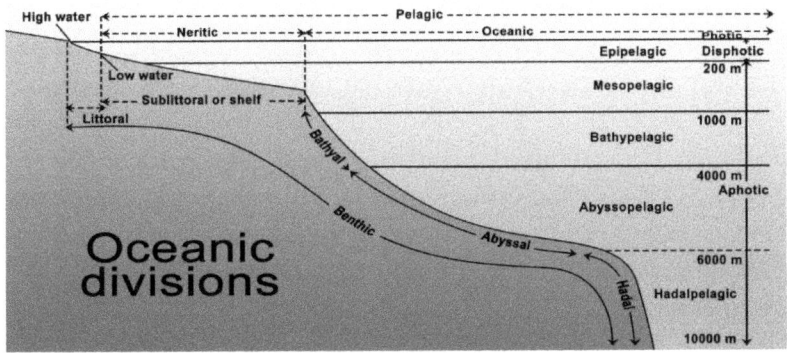

Image 1-2: Oceanic divisions

The ocean is divided into many zones based on its vertical (depth), and horizontal variations. Ocean itself is divided into benthic and pelagic zones which refer to ocean bottom and water itself, respectively.

- *Neretic zone*

It is also known as coastal waters or sublittoral zone. It is a shallow marine area expanded from mean low water down to 660-feet deep. The water temperature stays steady with an abundant amount of sunlight and oxygen. It has lower water pressure and moderate salinity. The zone is located on the continental shelf. This zone provides various resources for human consumption. Runoff and dumping of wastewater also deliver pollutants to the neritic area and this area is subsequently among the ocean's most polluted areas. It can be divided based on tide levels. The higher layer is known as the intertidal zone, encompassing the region from the wave splash zone to the high tide mark. This further can be divided into the following:

Supralittoral zone: The area is called the splash zone or spray zone. The area beyond the high tide mark receives only wave splash and seawater mist. The spray zone is not inundated by ocean water. Seawater enters into these elevated areas only during storm surges with the highest high tide.

Eulittoral zone: The eulittoral zone or mid-littoral zone includes the common intertidal zone and submersion by tides. The eulittoral zone expands from the spring high tide line, which is seldom submerged to the spring low tide line and is not inundated frequently. It is alternately exposed and submerged once or twice daily bases.

Subtidal zone: This zone is always covered with water. The coastal life that remains underwater. The subtidal zone is significant in tidal flow, energy dissipation, internal flows, river outflows, etc. this zone refers to areas where sunlight can reach the floor. In the subtidal zone, sufficient sunlight is found which can support primary production into the ocean. Thus, the majority of marine life is living in the subtidal zone.

- *Oceanic zone*

It refers to the area of the open sea beyond the edge of the continental shelf. The depth of the oceanic zone is beyond 200 meters; it is the area of open sea outside the edge of the continental shelf and includes 65% of the ocean's entirely open water. Large animals such as whales and colossal groupers may survive their full lives in the open water. The few species are live in the most depths are extremely specific. The oceanic zone can be divided into five-zone; Epipelagic zone, Mesopelagic zone, Bathypelagic zone, Abyssopelagic zone, and Hadalpelagic zone.

Epipelagic zone: It is the utmost layer of the ocean. The epipelagic zone is in depths of 200 meters in tropical and subtropical latitudes and about 100 meters in higher latitudes. In this layer all photosynthesis takes place. The epipelagic zone only represents 2-3% of the whole ocean, light is too dim for photosynthesis to occur. The epipelagic zone is copious in marine life than the lower zones; in particular, phytoplankton cannot live at any lower depths. In fact, in the entire world's oceans, only 65% of the plankton is in the top 500 meters.

Mesopelagic zone: The next zone encountered is the mesopelagic or twilight zone or disphotic zone. It is called twilight because of dimly light dispersion. In this zone, enough light penetration for visual predators to see, and even for a little photosynthesis. (Describe in chapter 3)

Bathypelagic zone: It starts at the bottom of the mesopelagic and extends down to 4000 meters. This zone is much larger than the mesopelagic zone and more than fifteen times the size of the epipelagic. Organisms are relying on detritus for food or on eating other animals in this zone. At this depth and pressure, generally, the fish, molluscs, crustaceans, and jellyfish are found. Sperm whales are hunt at these depths on the instance to prey on giant squid.

Abyssopelagic zone: It is also known as the abyssal zone. It stretches from the bottom of the bathypelagic to the seafloor. This zone is considered a comparative lack of life. It truly is the abyss. The abyssaopelagic zone is between 4000 meters to 6000 meters in depth. This zone is residue in everlasting darkness. It is covering regarding 83% of the total area of an ocean ecosystem and 60% of the terrestrial surface. The abyssopelagic zone contains a much higher concentration of nutrient salts, like nitrogen, phosphorus, and silica. Due to the great amount of dead organic material that drifts down from the beyond ocean zones and decay.

Hadalpelagic zone: This zone is also known as the Hadal zone or Aphotic zone. It is the deepest zone. The area is below the abyssal zone is the sparingly inhabited hadal zone. It is open waters of deep oceanic trenches between 6000 meters to 10000 meters. The increasing area occupied by the 46 different hadal habitats globally is less than 0.25 percent of the world's seafloor, however, trenches account for over 40 percent of the ocean's depth range. The deepest ocean trenches are the slightest explored and most tremendous of marine ecosystems. It has an absolute lack of sunlight, low temperatures, nutrient paucity, and extremely high hydrostatic pressures. Most organisms have scavengers and detrivores. (Described in chapter 3)

References:

- Beaverton, R.J.H. 1984. Dynamics of single species.: R.M. May (ed). The exploitation of Marine Communities. Berlin: Springer Verlag. pg. 13-58.
- Belzunce M.J., Saloun O., Valencia V. and Perez V. 2004: Contaminants in estuarine and coastal waters. In: Oceanography and Marine Environment of the Basque country. A. Borja and M. Collins (Eds.). Elsevier Oceanography Series. Amsterdam. 70, pg. 33- 51.
- Bouchet Ph., Lozouet P., Maestrati Ph., Heros V. 2002. Assessing the magnitude of species richness in tropical marine environments: exceptionally high numbers of molluscs at a New Caledonia site. Biological Journal of the Linnean Society, 75: pg. 421-436.
- Bristow, L.A., Dalsgaard, T., Tiano, L., Mills, D.B., Bertagnolli, A.D., Wright, J.J., Hallam, S.J., Ulloa, O., Canfield, D.E., Revsbech, N.P., and Thamdrup, B. 2016. Ammonium and nitrite oxidation at nanomolar oxygen concentrations in oxygen minimum zone waters. Proc. Natl. Acad. Sci. USA 113, pg. 10601–10606.
- Cave R., Ledouox R.I., Jickells K.T., Andrews J.E. and Davies H. 2003: The Humber catchment and its coastal areas: From UK to European perspectives. Science of Total Environment. pg. 314-316.
- Holbrook, W. S., Páramo, P., Pearse, S., and Schmitt, R.W.:2003. Thermohaline fine structure in an oceanographic front from seismic reflection profiling, Science, 301, pg. 821–824.

- Maria Byrne, 2011. Impact of ocean warming and ocean acidification on marine invertebrate life history stages: Vulnerabilities and potential for persistent in a changing ocean. Oceanography and Marine biology; An Annual Review, pg. 491-42.
- Moore, C.M., Mills, M.M., Arrigo, K.R., Berman Frank, I., Bopp, L., Boyd, P.W., Galbraith, E.D., Geider, R.J., Guieu, C., Jaccard, S.L., et al. 2013. Processes and patterns of oceanic nutrient limitation. National Geoscience 6, pg. 701–710.
- Norman L. Christensen., Ann M. Bartuska, James H. Brown, Stephen Carpenter, Carla D'Antonio, Rober Francis, Jerry F. Franklin, James A. MacMahon, Reed F. Noss, David J. Parsons, Charles H. Peterson, Monica G. Turner and Robert G. Woodmansee 1996. The report of the ecological society of America committee on the scientific basis for ecosystem management. Ecological Applications, 6(3): pg. 665-691.
- Portner and Knust 2007. Climate change affects marine fishes through the oxygen limitation of thermal tolerance. Science 315: pg. 95-97.
- Ray, G.C. 1988. Ecological diversity in coastal zones and oceans. In E.O. Willson (ed), Biodiversity National Academy Press, Washington, D.C. Pg. 36-50.
- Saito, M.A., Bertrand, E.M., Dutkiewicz, S., Bulygin, V.V., Moran, D.M., Monteiro, F.M., Follows, M.J., Valoia, F.W., and Waterbury, J.B. 2011. Iron conservation by reduction of metalloenzyme inventories in the marine diazotroph Crocosphaera watsonii. Proc. Natl. Acad.Sci.USA108, pg. 2184–2189.
- Van Mooy, B.A.S., Fredricks, H.F., Pedler, B.E., Dyhrman, S.T., Karl, D.M., Koblizek, M., Lomas, M.W., Mincer, T.J., Moore, L.R., Moutin, T., et al. 2009. Phytoplankton in the ocean use non-phosphorus lipids in response to phosphorus scarcity. Nature 458, pg. 69–72
- Van Syoc, R. J., Fernandes, J. N., Carrison, D. A. and Grosberg, R. K. 2010. Molecular phylogenetics and biogeography of

Pollicipes (Crustacea: Cirripedia), a Tethyan relict. Journal of Experimental Marine Biology and Ecology, 392.

- Vergés, A., Steinberg, P. D., Hay, M. E. et al. 2014. The tropicalization of temperate marine ecosystems: climate mediated changes in herbivory and community phase shifts. Proceedings of the Royal Society B: Biological Sciences, 281.
- Vieira, R., Pinto, I. S. and Arenas, F. 2017. The role of nutrient enrichment in the invasion process in intertidal rockpools. Hydrobiologia, 797, pg. 183–98.
- Viejo, R. M., Arenas, F., Fernández, C. and Gómez, M. 2008. Mechanisms of succession along the emersion gradient in intertidal rocky shore assemblages. Oikos, 117, pg. 376–89.
- Vinagre, C., Mendonça, V., Narciso, L. and Madeira, C. 2015. Food web of the intertidal rocky shore of the west Portuguese coast – determined by stable isotope analysis. Marine Environmental Research, 110, pg. 53–60.

Chapter 2
Oceans of the World

The ocean is allied to key lakes, watersheds, and waterways because the entire foremost watersheds on earth are conduits to the ocean. So no matter where you reside you are linked to the world's large universal ocean! The ocean's depth is fixed, and resources are limited. It can be found in a variety of forms, including rivers, estuaries, lakes, the sea, and ice. The changing sea level has extended and contracted continental shelves, formed and smashed inland seas, and shaped the terrain's surface. Oceanic water accounts for approximately 97 percent of the hydrosphere. The remaining three percent is ice-covered and distributed as icecaps in lakes, rivers, subsurface aquifers, and water vapour.

The sea can consider under the ocean, but the ocean cannot consider under sea. There are five oceans in the world; the Pacific Ocean, Atlantic Ocean, Indian Ocean, Southern Ocean, and Arctic Ocean. The Great Lakes are an example of an inner sea that has been shaped by more than 3 billion years of erosion, marine sea flooding, sedimentation, and glaciations. The oceans dominated earth's carbon cycle. Half of all primary productivity on earth receives a place in the sun-drenched layers of the ocean, which absorbs roughly half of all carbon dioxide added to the environment. The ocean influences areas of the world that receive additional or less rainfall. The initial amount of rain that falls on the ground.

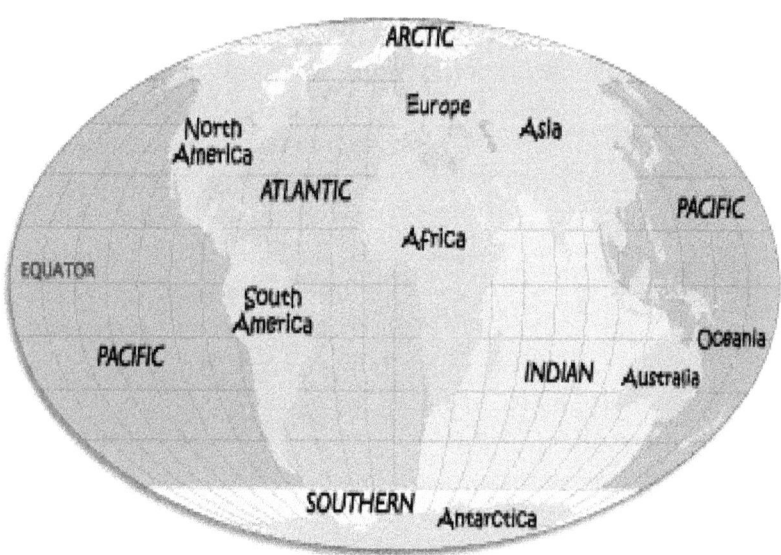

Image 2-1: Map of the world ocean

2.1 The Pacific Ocean

Its name was given by Ferdinand. The word 'Pacific' connotes a peaceful world in Portuguese. It's larger than the earth's whole landmass. Geographically, it alone accounts for approximately 32% of the Earth's surface. There are over 25,000 islands in the Pacific Ocean. It is divided into two sections: north and south pacific. North Pacific latitude meets European latitude, while South Pacific latitude meets America, Africa, and Australia. It is close to many ports and harbors. It stretches from North and South America to Asia and Australia.

The Pacific Ocean is surrounded by the Ring of Fire, a chain of volcanoes. It also serves as a habitat for the Great Barrier Reef, Australia's largest coral reef. Around the South Pacific, three plate tectonics are interacting, resulting in earthquakes and volcanic eruptions. It travels along the coasts of North and South America, Asia, and New Zealand. The Mariana Trench in the Pacific Ocean is the world's deepest point, reaching a depth of 10,911 meters.

The Great Barrier Reef in Australia is the world's longest reef. The Pacific Ocean is teeming with life. Deep-sea life forms evolve as a result of the activation of volcanoes. Phytoplankton blooms provide fresh air on the planet. The Pacific Ocean is approximately 60,000 square miles in size. On the Pacific Ocean, scientists had made a magnificent discovery. They found in seafloor vents gushing Shimmering, warm mineral-rich fluids into

---

Did You Know...?

- The pacific ocean has the world's predominant, innermost, oldest, widest, and most seismically active body of water.
- The Ring of Fire is more than 40,200 km long chain of tectonic activity, including earthquakes, volcanoes, mountains, islands etc.
- The Ring of Fire is home to 75% of all active volcanoes in the world.

the cold, dark depths. It surprised to see life in the hydrothermal vent. The Pacific Ocean is entirely enclosed by subduction zones. It produces immense seismic and highly active volcanoes.

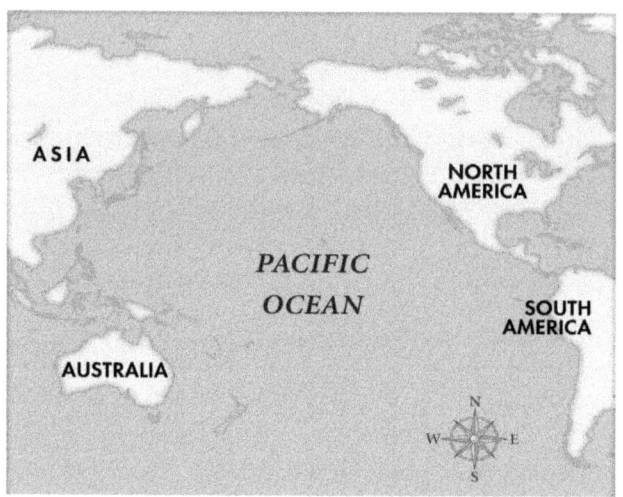

Image 2-2: Region of the Pacific Ocean
(Source: The encyclopedia Britanica, Inc.2012)

- *Ring of Fire*

There are over 350 historically active volcanoes in the ring of fire. It's worth noting that the Ring of Fire is home to 90% of the world's earthquakes. Historically around 89% of the world's largest earthquakes have occurred outside of a ring of fire. It is linked to a nearly continuous succession of trenches, volcanic arcs and volcanic belts, plate movement, and so on. Four continents (Asia, Oceania, North America, and South America) and eight tectonic plates (the Pacific plate, the Cocos plate, the Nazca plate, the Philippine plate, the North American plate, the Eurasian plate, the Indo-Australia plate, and the South American plate) are affected by the ring of fire. Subduction occurs when a fast-moving plate is subducted beneath a slower-moving plate, such as the Philippine plate. Volcano eruptions can be beneficial in other ways, such as valuable mine deposits, fertile soils, geothermal energy, and the formation of new islands, such as Hawai Island. It spans a zonal distance of 20,000 kilometers from Malacca Strait to Panama in the tropics. The term "volcano" is derived from the Latin word "Vulcan" in Greek mythology. It is believed through this pathway an opening to the fiery underworld. Hephaestus, the god of fire, lived beneath Mt.Etna(Sicily, Italy). It is a guard volcano formed by a series of lava flows. Shield volcanoes are great volcanoes that are built regarding wholly of fluid lava flows, have broad sloping sides, and are generally enclosed by gently sloping hills in a circular or fan-shaped outline that looks like a warrior's shield. It has no other layers, as the volcano just consists of lava.

Image 2-3: One of the most active volcanoes -Mt.Etna, Italy
(Source: BBC)

- *Great Barrier Reef*

The Great Barrier Reef is the largest reef system in the world. It is located on the northeast coast of Queensland, Australia is a place of incredible variety and splendor. It is covering over 900 islands. Reef lays in the top northeast area of Australia. It is parallel to a landmass, which is typically several miles away. The Great Barrier Reef is the world's only structure which is made up of living organisms. It can be visible from outer space. It is a structure that dates back more than 500,000 years old. Climate changes, toxic spills, overfishing, pouching, physical damage, and water quality are major issues for Great Barrier Reef pollution. The reef is home to over 1,600 species of fish, 450 species of hard coral, and 150 species of soft coral, as well as an enormous diversity of sponges, anemones, marine worms, crustaceans, and other species. The reef is also home to whales, dolphins, rays, seahorses, sharks, sea snakes, shellfish, and birds. Furthermore, it is of great scientific importance as the habitat of species such as the dugong and the great green turtle. These animals help to make the great barrier reef one of the world's richest and most diverse natural ecosystems, as well as a haven for many species of conservation concern.

- *Mariana Trench*

It is also referred to as Challenger Deep. The British vessel HMS Challenger 2 is the first to explore the Mariana trench. It is east of the Mariana Islands and west of the Pacific Ocean. It is the world's deepest trench. The mariana trench is about 1580 miles long and 43 miles wide. It is incredibly cold and there's a lot of pressure. These incredible pressures must be tolerable for the deep-sea fauna. Goblin Shark, Hatchet Fish, Footballfish, Frilled Shark, Dumbo Octopus, Ping-Pong Tree Songe, and Enypniastes are some examples. they have different proteins that are adapted for life under these conditions. Deep diver organisms live a long time. they can survive for over a century. They have slow developmental processes. The Philippine plate subducted and submerged the fast-moving

Pacific plate. Crust from the eastern Pacific ridge parallel to the mariana trench was used to build the mariana trench. At the Mariana trench, the new crust formed by the east Pacific ridge moved and subducted. Some researchers have found life in the depth in which inhabitants look terrible- octopus, unusual sea stars, and many inimitable creatures.

Table 2-1: Ports and Harbours at the Pacific Ocean

Region	Country	Region	Country
Central America	El Salvador, Sonsonate	Oceania	New Zealand
	Nicarahua		United States, Hawaii
	El Salvador, La Liberted		Australia, Victoria
	Panama		Australia, New South Wales
	Costa Rica		Australia, Tasmania
	Honduras		Australia, Northern Territory
	Guatemala		Australia, Queensland
North America	Mexico, Guerrero		Australia, Papua New Guinea
	Mexico, Baja California Sur	Southeast Asia	Philippines, Negros
	U.S., California		Philippines, Albay
	Mexico, Sinaloa		Thailand
	U.S., Washington		Vietnam
	U.S., Alaska		Indonesia, North Sulawesi
	Canada, British Columbia		Indonesia, East Kalimantan

Region	Country	Region	Country
	U.S., Oregon		Malaysia, Pahang
	Mexico		Malaysia, Sabah
	Mexico, Sonora		Malaysia, Terengganu
	Mexico, Jalisco		Malaysia, Sabah, Kuala Penyu
	Mexico, Colima		Malaysia, Sarawak
	Mexico, Michoacan		Singapore
	Mexico, Oaxaca		Brunei
	Mexico, Chiapas		Cambodia
South America	Chile	East Asia	South Korea
	Colombia, Valle del Cauca		North Korea
	Peru		China, Guangdong
	Ecuador		China, Hongkong
	Colombia, Narino		Taiwan
	Argentina		Japan
			Russia

2.2 The Atlantic Ocean

The Atlantic Ocean is a relatively small body of water. It is the second-largest ocean in terms of area after the Pacific Ocean. It is thought to have formed during the Jurassic period. It covers roughly one-fifth of the earth's surface. The Atlantic Ocean was first mentioned in the histories of Herodotus in 450 BC as Atlantis Thalassa. The name of the ocean is derived from Greek mythology and means "Sea of Atlas". It exists in the space between two parallel continental masses. On the eastern side, it is bordered by Europe and Africa. It is surrounded by North and South America, as well as its western region. Scientists frequently divide the Atlantic into two basins: the north and south america. Aside from the Baltic Sea, several ports and harbors are located near the Atlantic Ocean. The North Atlantic, where waters are submerged after being chilled by arctic temperatures, marks the beginning of the "global ocean conveyor," a motion pattern that aids in climate control. The Atlantic has no distinct northern or southern boundaries. It flows north into the Arctic Ocean and south into the Antarctic Ocean. The ocean's shape is generally interrupted, S-shaped, and slender. The Atlantic Ocean's area without its dependent seas is approximately 31,830,000 square miles, and its area among them is approximately 41,100,000 square miles.

It is located in the Puerto Rico Trench, north of the island of Puerto Rico, and has an average depth of 3,300 meters and a maximum depth of 8,380 meters. A glimpse at the distribution of high-quality ocean data tells that the Atlantic Ocean is by far the well-researched part of the world ocean. This is especially true for the North Atlantic Ocean, which is home to many oceanographic research institutions in the United States and Europe. It is remarkable to find that the world's second-largest ocean has by far the largest drainage area. It receives a large portion of the major rivers such as

The Amazon, Black, Congo, Elbe, Mississippi, Niger, Orinoco, Rio de la Plata, St. Lawrence, and others, and drains into the Black, Baltic, and Mediterranean Seas. Pollution concerns in the Atlantic Ocean include agricultural and municipal waste. Eastern southern Brazil, the United States, and eastern Argentina contribute to metropolitan pollution. Oil pollution is a problem in the Caribbean Sea, the Gulf of Mexico, the Mediterranean Sea, and the North Sea. It is home to a variety of species, including sea turtles and dolphins. Those that we can see on the surface as well as those that are hidden from human eyes. Furthermore, it is a spawning ground for both European and American eels. National Geographic published photos of various corals, mollusks, fishes, and crabs in 2018. The North Atlantic right whale is on the verge of extinction. Only 400 people remain in the wilderness. The term "right" was coined by early-twentieth-century whalers who believed it was the "right" whale to catch.

Image 2-4: Region of the Atlantic Ocean

- **Bermuda**

It is located in the North Atlantic Ocean. It is also known as a 'Devil Triangle'. It is connecting Triangle to Florida, Puerto Rico, and Bermuda. It is a place where several airplanes and ships have mysteriously vanished under the mysterious Triangle. Approximately 2000 ships and 75 airplanes vanished here. The Atlantic is one ravenous Ocean. It is nourished by an inconsistent number of the world's major rivers and about half of the world's landmass. Just a few of these rivers include Mississippi, Amazon, Congo, Niger, Loire, Rhine, and great rivers draining into the Mediterranean and Baltic seas. The Atlantic has the utmost tides of any ocean.

> **Did You Know..?**
> - The North Atlantic is rich in islands, in the range of its coastline and tributary seas.
> - It's the second largest ocean. Covering 22% of the earth.
> - The name is derived from legendary island of Atlantis as described by Plato.

The temperatures are depending on the location and the ocean's currents. The closer to the equator the warm the water be likely to be. The higher temperature of 28°C is reached in coastal regions near the equator and the minimum temperatures are around -2°C in the polar regions.

The Atlantic Ocean contains several islands, the most well-known of which are The Bahamas, Canary Islands (Spain), Azores (Portugal), Cap Verde Islands, and Greenland. Greenland is the largest island in the world. The straits of Gibraltar, which connect Spain and Morocco, are important Atlantic Ocean waterways. The Bosporus in Turkey is one of the Atlantic Ocean's most important waterways. The longest stretch of ocean is 2,848 kilometers between Brazil and Sierra Leone.

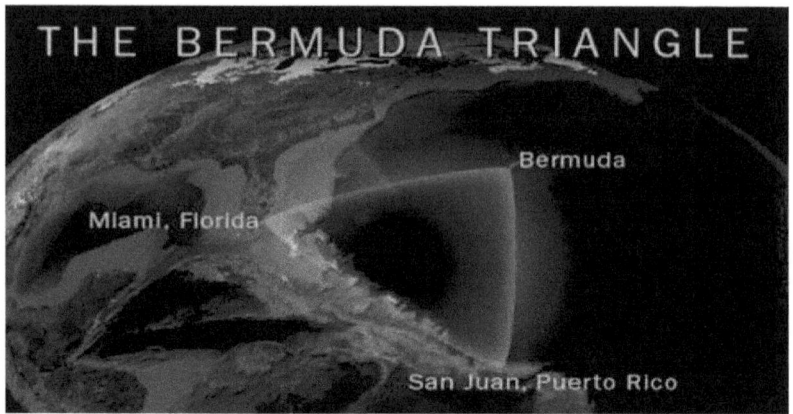

Image 2-5: The Bermuda Triangle (Source: NOAA)

- *RMS Titanic*

Great Britain, Cuba, Ireland, Great Britain, and Newfoundland are the largest islands in the Atlantic. It is very interesting to know that in some areas water levels have risen more than fifty-three feet. Icebergs in the Atlantic Ocean are causing many ships to sink. At the time she entered service, the RMS Titanic was the largest ship buoyant. The Titanic is one of the most famous shipwrecks in history. The RMS Titanic sinks in the North Atlantic Ocean 400 miles off the coast of Newfoundland, Canada, on April 15, 1912. During her maiden voyage from South Hampton to New York City, she was hit by an iceberg and sank. More than 2224 passengers and crew traveled on the ship; approximately 1500 of these passengers died. It was carried by several of the world's wealthiest people.

It is the first time an airplane has flown across an ocean. The Cunard Line was the first ship to cross the Atlantic Ocean with passengers on board in 1850. It was also the first ocean crossed by a ship. It is the hometown of the second-largest barrier reef in the world, the Cancun Reef off the coast of

Mexico. The Atlantic ridge is a submerged mountain range located in the North Atlantic Ocean. Diamonds can be found seabed off the coast of southern Africa. It is the house of various fauna like a manatee, humpback whale sea, starfish, catfish, atlantic ghost crab, penguins, and so on.

Table 2.2: List of ports and harbors at the Atlantic Ocean

Region	Country/Territory	Region	Country/Territory
Africa	Ghana	South America	Argentina
	The Gambia		Falkland Islands
	Guinea		Guyana
	Senegal		Uruguay, Rocha
	Cameroon		Uruguay
	Togo	South America-Brazil	Amapa
	Angola		Para
	Equatorial Guinea		Ceara
	Liberia		Rio Grande do Norte
	Cape Verde		Paraiba
	Algeria		Pernambuco
	Tunisia		Alagos
	Namibia		Sergipe
	Sierra Leone		Bahaia
	Benin		Espirito Santo
	Cote d Ivoire		Rio de Janeiro
	Gabon		Sao Paulo
	Mauritania		Parana

Region	Country	Region	Country
	Morocco		Santa Catarina
	Nigeria	South America-Chile	Magallanes
	Halena, Ascension and Tristan da Cunha		The Chilean Antarctica
	South Africa	South America-Colombia	Atlantico
	Democratic Republic of Congo		Bolivar
	Egypt		Magdalena
Africa/ European Union	Canary Islands		La Guajira
North America-Canada	Prince Edward Island,		Antioquila
	Newfoundland and Labrador	South America / European Union	French Guiana
	Nova Scotia	Mid-Atlantic/ South America	South Georgia
	New Brunswick		The South Sandwich
	Quebec		Suriname
North America-Mexico	Tamaulipas	South America-Venezuela	Bolivar
	Yucatan		Anzoategui
	Veracruz		Vargas
	Quintana Roo		Zulia

Region	Country	Region	Country
	Tabasco		Nueva Esparta
	Campeche		Carabobo
	Michoacan	Caribbean	Puerto Rico
North America-United States	New York		Curacao
	Virginia		Saint Kitts and Nevis
	Florida		Cuba
	Georgia		Jamaica
	Maryland		Barbados, Saint Michael
	Maine		St. Vincent
	Louisiana		Bahamas
	Texas		Guadeloupe
	Massachusetts		Dominican Republic
	Connecticut		Martinique
	New Jersey		Aruba
	South Carolina		Haiti
	Mississippi		Trinidad
	Alabama		Tobago
	Rhode Island		Honduras
	Pennsylvania		Guatemala
	New Hampshire	Europe	Belgium
	Trenton		Estonia
	District of Columbia		France
	Delaware		Germany
	North Carolina		Spain

Region	Country	Region	Country
Mid-Atlantic/ Europe	Faroe Islands		Cantabria
	Iceland		Italy
	Netherland		Denmark
	Norway		Sweden
	Portugal		Montengro
	Russia		Albania
	Spain		Slovenia
Europe- United Kingdom	Wales		Crotia
	Northern Ireland		Greece
	England		Russia
	Scotland		Romania
Central America	Honduras		Ukraine
	Costa Rica, Limon Province		Guernsey
	Panama		Isle of Man
	Guatemala		Ireland
	Belize		Gibraltar
West Asia	Turkey		Malta
	Georgia		Netherland
	Israel		
	Lebanon		
	Cyprus		
	Syria		
	Sicily-Italy		
	Norway		

2.3 The Indian Ocean

The Indian Ocean is the third-largest ocean in the world. It covers 20% of the world's land surface. The ocean's average depth is 3890 meters. The name comes from India. It is also referred to as 'Ratnakar.' It is the warmest Ocean in the world and also has a closed ocean. The area is approximately 73.6 million square kilometers. Diamantina is the deepest point in the Indian Ocean. It is the most recent ocean; having formed 100 million years ago. The Indian Ocean is bounded by the continents of Africa, Asia, and Australia. Around the Indian Ocean, well-known civilizations such as Sumerian, Egyptian, and the Indian valley civilization arose. It is bounded on the west by Africa and the east by Australia and Indonesia. Asia is to the north, while Antarctica is to the south.

The Indian Ocean is enclosed on the north by India, Iran, Pakistan, and Bangladesh. The Indian Ocean's western margin is separated from the Atlantic Ocean by the Suez Canal and a meridian running south from Cape Agulhas in South Africa. It touches the Pacific Ocean at the 147°E meridian, which runs south from Tasmania's South East Cape to 60°S latitude. The temperature of the Indian Ocean varies according to location and ocean currents. The warmer the water, the closer it is to the Equator.

The Indian Ocean is made up of several seas: The Andaman Sea, the Arabian Sea, the Bay of Bengal, the Great Australian Bight, the Gulf of Aden, the Gulf of Mannar, the Gulf of Oman, the Laccadive Sea, the Mozambique Channel, the Persian Gulf, and the Red Sea. The Indian Ocean Region is bounded and influenced by many countries, including Australia, Bahrain, Bangladesh, Comoros, Djibouti, East Timor, Egypt, Eritrea, India, Indonesia, Iran, Iraq, Israel, Jordan, Kenya, Madagascar, Malaysia, Maldives, Mauritius, Myanmar, Oman, Pakistan, Qatar, Saudi Arabia, Seychelles, Singapore, Somalia, South Africa, Sri Lanka, Sudan, Thailand, United Arab Emirates,

Yemen, and others. Australia, Indonesia, India, Madagascar, Malaysia, Thailand, Somalia, South Africa, and Saudi Arabia lead the way. The majority of the islands in this ocean are separated by continental drift, including the Andaman and Nicobar Islands, Sri Lanka, Madagascar, and Zanzibar. The Indian Ocean meets the Pacific and Atlantic oceans near the continent of Antarctica. Notably, the water of the Pacific and Indian Oceans meet but do not mix. Although the water of oceans flows in only one direction, the Indian Ocean's water flows towards India in the summer and towards Africa in the winter. Two of the world's largest rivers, the Brahmaputra and the Ganges, are submerged in this ocean water body. This ocean has many broad submerged ridges and basins. It has an average water mass of 292,131,000 cub/km. There are numerous prominent islands in the Indian Ocean, including Mauritius, Seychelles, Reunion, Madagascar, The Comores, Maldives (Portugal), Sri Lanka, and others. It is made up of the center of the major Indo-Pacific warm pool, which is mixed with the atmosphere and influences the climate both regionally and globally.

The thermocline is responsible for the world's strongest monsoon. This results in significant seasonal variations in ocean currents, such as the reversal of the Somali Current and the Indian monsoon current. The vicious monsoon brings rain to Bangladesh, Bhutan, India, Maldives, Nepal, Pakistan, and Sri Lanka, among other Indian subcontinents. Although winds are milder in the southern hemisphere, except for Mauritius, they can be severe. While wind patterns change, heavy cyclones smack the coasts of the Arabian Sea and the Bay of Bengal.

Fascinating detail from india is the discovery of a submerged continent in this ocean known as the 'Kerguelen Plateau,' which is thought to have volcanic origins. It is located in an Antarctic plate in the southern ocean. It is the second-largest oceanic plate after Ontong-Java in the Pacific Ocean.

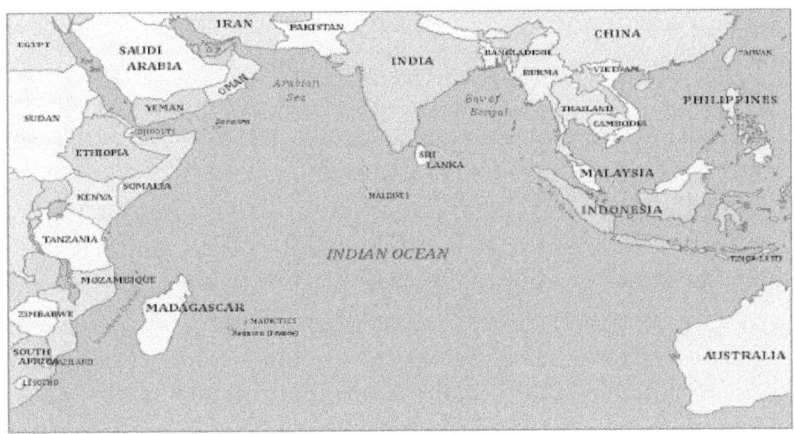

Image 2-6: The surrounding area of Indian Ocean margins

One of the busiest transportation routes in the Indian Ocean. The two most well-known waterways in the Indian Ocean are the Suez Canal in Egypt and the Malacca Strait between Malaysia and Indonesia. The Suez Canal was built to connect the Red Sea and the Mediterranean Sea. The northern Indian Ocean is also the most important oil transport route, connecting the oil-rich nations of the Middle East with Asia. Every day, tankers transport approximately 17 million barrels of crude oil from the Persian Gulf.

- *Tsunami*

The 26th of December, 2004 is remembered as a black day in the Indian Ocean. It is also well-known as the Boxing Day tsunami. A geological fault burst occurred off the west coast of northern Sumatra, Indonesia. A Tsunami struck the countries surrounded by the Indian Ocean. The epicenter was located at 3.31°N and 95.95°E, about 250 km south-southeast of banda Aceh, the capital city of Aceh, Northern Sumatra, Indonesia, The rupture continued to northward for more than 745 miles.

It causes an enormous 9.3 Magnitude earthquake to occur for 8-10 minutes. More than 26,000 people died as a result, and one million people were displaced. The majority of the fatalities occurred in Indonesia. It was intense and destroyed almost all villages, towns, roads, buildings, etc.

The Indus and Ganges rivers cover 40% of the Indian ocean's sediment. The polar front has high biological productivity and siliceous ooze. During the summer, the Indian Ocean has one of the highest concentrations of phytoplankton blooms. The monsoon season has a significant impact on a water mixture (upwelling). Upwelling helps nutrients into the upper zones where there is sufficient light. This ocean is home to a plethora of islands. These islands formed in various ways many years ago. Madagascar, located off the east coast of southern Africa. It is the largest island. It used to be connected to continents. Volcanoes are the foundation of Sumatra and reunion.

The well-known Maldives also formed beside coral atolls. These are flags of coral reefs that are produced around the tops of submerged volcanoes. Waste from several factories and cities along the coasts has polluted the waters. In addition, there is concern about the quantity of oil being transported in ships on the ocean.

> **Did You Know..?**
>
> - This ocean has a higher rate of evaporation.
> - The natural resources include oil, gas, heavy metal, natural gas, etc.
> - Allied and Axis navies fought many battles during the second world war-II.

The biodiversity of the Indian ocean is poorly known; it is used as a sort of proxy for areas known to have high biodiversity. Typically, research data is only available for groups with broad human interests, such as fish, corals, marine mammals, sea turtles, cephalopods, seagrass, and mangroves.

Still, there is a lack of data, or more specifically, a lack of comprehensive data for many countries.

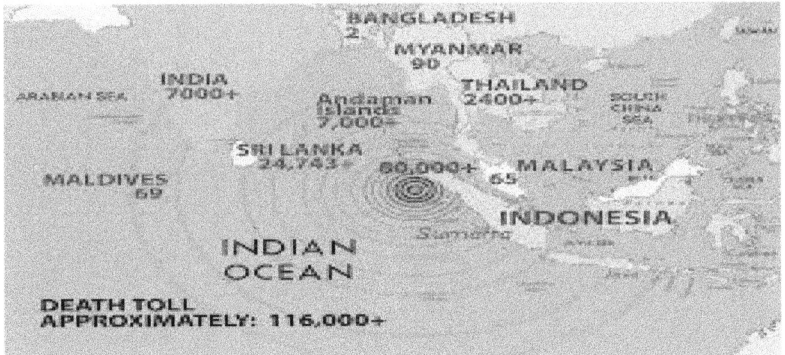

Image 2-7: Tsunami causes countries (Sources: Pacific disaster management information network)

Table 2-3: List of Ports and Harbours at Indian ocean

Region	Country/Territory	Region	Country/Territory
Oceania	Western Australia	Southern Africa	South Africa
	Tasmania	Southeast Asia	Myanmar
	Victoria		Malaysia
South Asia	Bangladesh, Chittagong	Southeast Asia/Isular	Maldives
	Bangladesh, Bagerhat	East Africa/Isular	Mauritius
	Bangladesh, Patuakhali		Seychelles
	Bangladesh, Cox's Bazar		Madagascar
	India, Gujarat	West Asia	Iran

Region	Country	Region	Country
	India, Maharashtra		Iraq
	India, Kerala		Israel
	India, Tamil Nadu		Oman
	India, Andaman and Nicobar Islands		Qatar
	India, Andhrapradesh		Saudi Arabia
	India, Odisha		United Arab Emirates
	India, Goa, Lakshadweep		Yemen
	India, Puducherry		Saudi Arabia
	India, West Bengal		Bahrin, Al Hidd, Manama
	India, Kerala	Northeast Africa	Egypt
	Pakistan, Baluchistan	West Asia	Eritriea
	Pakistan, Sindh	East Africa	Eritriea
	Pakistan, Balochistan		Kenya
	Singapore		Kuwait
	Sri Lanka		Mozambique
Southeast Asia	Indonesia, Bali, Riau		Somalia
	Indonesia, North Sumatra		Somaliland
	Indonesia, Central Java		Tanzania
	Indonesia, Jakarta		Djibouti
	Indonesia, Surabaya		Myanmar

2.4 The Southern Ocean

The Great Southern Ocean, Antarctic Ocean, South Polar Ocean, and Austral Ocean are all names for the southern ocean. It covers approximately 6% of the earth's surface. It was previously known as the Antarctic Ocean. However, there is some debate over whether the Earth has four or five oceans. The International Hydrographic Organization (IHO) issued a report in 2000 that confirmed the Southern Ocean as a new ocean. Simultaneously, the IHO asked its sixty-eight-member nations to vote on a name for the new ocean. The majority of the 28 countries that voted chose the Southern Ocean. As a result, many people still refer to it as the Antarctic Ocean, although the official name is the Southern Ocean. Antarctica is almost entirely covered in ice.

The Atlantic name means 'Opposite to the North' in Greek. The South Pole is surrounded by the Southern Ocean. The Southern Ocean is the youngest, having been twisted only 30 million years ago when South America and Antarctica separated. It is the world's fourth-largest ocean. The southern ocean can be found in the southern hemisphere. It is encircled by Antarctica. Antarctica's neighboring continents are South America, New Zealand, Australia, and Africa. As a result, the southern ocean is said to 'emerge' from the water of the South Atlantic Ocean.

Both the South Pacific and Indian Oceans congregate near latitudes south of 60°south. It covers 4,479,000 square miles and has a capacity of 17, 226,000 cubic miles. The depth of the southern ocean ranges from 12,000 ft to 14,000 ft., with an average depth of 10,700 ft. The South Sandwich Trench, located southeast of the island of South Georgia, has a depth of over 23,000 feet.

- *Antarctic Circumpolar Current*

The Atlantic Current is the world's largest ocean current. From west to east, it transports 135 to 145 million cubic meters of water per second. The sea temperature ranges from -2°C to 10°C. The sea temperature is below zero degrees during the winter season (April to October).

The region's low temperatures, however, do not deter tourists who are on a special adventure of travels and expeditions. Every year, over 50,000 tourists visit the Southern Ocean.

The southern ocean is distinguished by its striking diversity. It is the largest home of 'Blue Whale'. The largest and deepest invertebrate is the colossal squid. It has the largest eye in the world. The wandering albatross has the world's largest wingspan. The Antarctic region also serves as a biological boundary. Many organisms live within areas, but they are extremely rare on the other side of the globe. The Southern Ocean's fishing activities are strictly regulated. Krill predominates in the catch. It has been subject to rapid climate change over the past 30 years, which has directly to changes in the marine ecosystem.

Antarctica has no indigenous people. There are only research stations from various countries on the coldest continent, but these researchers, scientists, and explorers, as well as their families, only work and live there for a partial time. There are only two communities, where people are live and work the whole year. It has fewer than 100 people in winter. Two research stations are located in the southern ocean: Chilean research station 'Villas Las Estrellas' and the Argentinian research station 'Esperanza Base'. Although being located in one of the main remote regions of the earth, it has only some ports that have occurred. Ports are mostly associated with research stations, such as the Rothera Station, Palmer Station, Mawson Station. However, the southernmost ports in Australia have been recognized as well as ports of the Southern Ocean.

Antarctica is thought to be a desert due to the fact that very little moisture falls on its surface. The Sahara desert gets more rain than Antarctica. Most of its moisture falls in the type of snow. It is believed that if the ice sheets in the Southern Ocean was to melt the oceans around the world would rise by as much as 65 meters. The ozone causes solar UV-light radiation.

> **Did You Know..?**
>
> - It only ocean that goes around the world.
> - Bartolomeo Dias, Portuguese explorer was the first discover the southern ocean(1487).
> - Armed forces activities are restricted.

The primary productivity of phytoplankton has decreased by about 15% in recent years, which is concerning because phytoplankton is the primary food source for krill, a key species in the Southern Ocean ecosystem. Recent studies on Antarctica have found that ultraviolet radiation from the sun has increased in recent years, resulting in a decrease in phytoplankton growth. The increased solar radiation emitted by Antarctica's ozone hole has also resulted in the modification of DNA in some fish species. The majority of Antarctic ports are run by government research stations and are not accessible to commercial or private ships. It has some of the strongest winds on earth.

Contaminants from coastal stations are introduced into some parts of the southern ocean via wastewater, dump sites, and other materials. Organic pollutants have been found in the water, sediments, and organisms in the vicinity of some stations regularly. Large-scale contamination spreads through the ocean's global circulation. Many pollutants are transported and biomagnified in the southern ocean, including DDT (dichlorodiphenyltrichloroethane) and some organophosphates. The snow samples allowed for the restoration of Antarctica's lead pollution, which had begun as early as 1880. As a result of long-distance transportation from the surrounding continents, the area is heavily contaminated with other metals such as chromium, copper, zinc, silver, bismuth, and uranium.

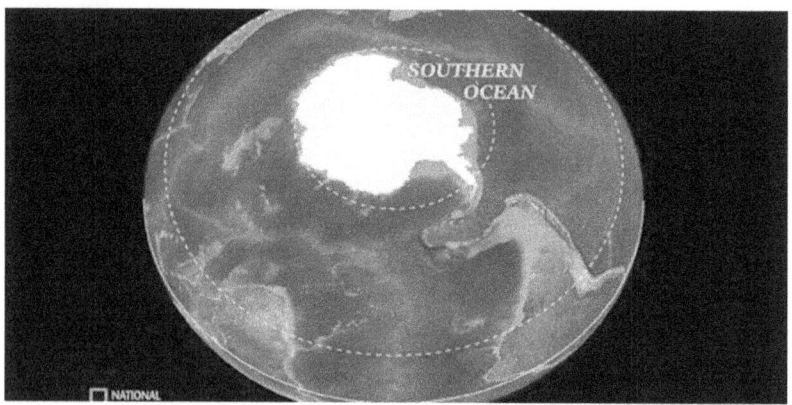

Image 2-8: The Southern ocean area (Sources; National Geographic)

Table 2-4: List of Ports and Harbours at Southern ocean

Region	Country/Territory	Region	Country/Territory
Antarctica	Adelie land	Antarctica	The Ross dependency
	Bouvet Island		The South Orkney Islands
	Graham Land		The South Shetland Islands
	Mac. Robertson Land		Wilkes Land
	Queen Maud Land		

2.5 The Arctic Ocean

The Arctic Ocean is one of the earth's last unexplored territories. It is becoming more reachable as global temperatures rise and Arctic sea ice declines. As global temperatures rise and Arctic sea ice melts, it becomes more accessible. The Arctic Ocean is located in the northern hemisphere, located north of 60°N latitude, and is bounded by the Eurasian and North American continents as well as numerous islands. On the earth, it covers only 3% of the earth's surface. The Arctic Ocean occupies roughly areas of about 14.1 million sq. km. The majority of the ocean is enclosed beside ice, but the thicknesses of the ice diverge depending on the season. Huge swaths of the Arctic Ocean are either seasonally or permanently encased in ice. Because of its low evaporation rates, it has the lowest salinity of the world's oceans. It's a frozen sea that's surrounded by land. The ocean also receives a large amount of fresh water from the numerous rivers that drain into it. The ecosystem in the area is fragile and deteriorating. Some areas of Greenland were exposed to the open ocean for the first time in millennia during 2018, the second-worst sea ice decline in history.

The Arctic Ocean's temperatures are moderately stable, hovering around -20°C all year. The weather changes with the seasons, and the sky is mostly overcast over the Arctic Ocean. From September to May, the winter season is in effect. Wind and ocean currents have an impact on sea ice. For more than two years, the soil in the arctic regions freezes! The Arctic ice is melting as a result of rising ocean temperatures caused by global warming. Every year, more ice melts in the summer, and less water freezes in the winter. The region's low temperatures, however, do not deter tourists who visit on unique adventure travels and expeditions. Every year, a large number of tourists visit the Arctic region. Most trips are started from Svalbard Island in Norway or Nuuk in Greenland. It has the lowest

Salinity among all of the oceans, due to Low evaporation, freshwater inflow from rivers and streams. The ocean is covered with polar ice, fast ice, and pack ice.

The Arctic Ocean has an average depth of 3406 feet. Litke deep is the deepest point in the ocean, measuring 17,880 feet. It has deep trenches and ridges, the largest of which is Lomonosov Ridge, which divides the ocean into the Amerasian and Eurasian basins. The Arctic ocean was once encased in a difficult mass of ice, posing a severe challenge to shipping. The sea is warming and expanding as a result of global warming. On the Arctic Ocean, races are beginning to take control of the land. The United States and Canada both have armed forces in the Arctic, and China has recently expressed interest in expanding its influence there. Russia and Norway, on the other hand have done the most to prepare their militaries and industries for safer Arctic Ocean waters. Several countries in the region are beginning to expand their oil and natural gas operations. It is acquiring a significant amount of natural resources such as petrol, natural gas, placer deposits, sand, fish, and so on. significantly, it holds 25% or more of the world's undiscovered oil and gas resources. The Arctic Ocean covers up the northern polar region of the Earth. The north pole is roughly in the center of it. The Arctic Ocean is surrounded by Russia, Finland, Sweden, Canada, Greenland, Iceland, Norway, Scandinavia (northern), and the United States (Alaska).

It is home to a variety of polar animals including whales, seals, walruses, and polar bears. seals are slaughtered for their fur and food. In general, it is exported to China and Russia for use in clothing. Seal meat is mostly exported to Asian countries. It is not easy to get to the Arctic ocean. The majority of the ocean here is dark, as sunlight is blocked by ice cover. There isn't enough information available about the sea's food chain. Plankton is microscopic organisms such as algae and bacteria that form the foundation of the Arctic food chain. They convert carbon dioxide from the atmosphere into organic matter, which feeds everything from small fish to large Bowhead whales.

Did You Know..?

- The arctic name is derived from greek word 'Arktos' which mean 'bear'.
- It is smallest and shallowest ocean in the world.
- Most scientist describe the arctic as the area above 'Arctic Circle'

Walrus have huge tusks that they use to Pull themselves along for hours at a time while swimming. Whales and fish are often a vital food source for indigenous people in the Arctic, but commercial fishing has been banned in much of the Arctic Ocean. Oceans serve as the planet's largest habitat and also help to regulate the global climate.

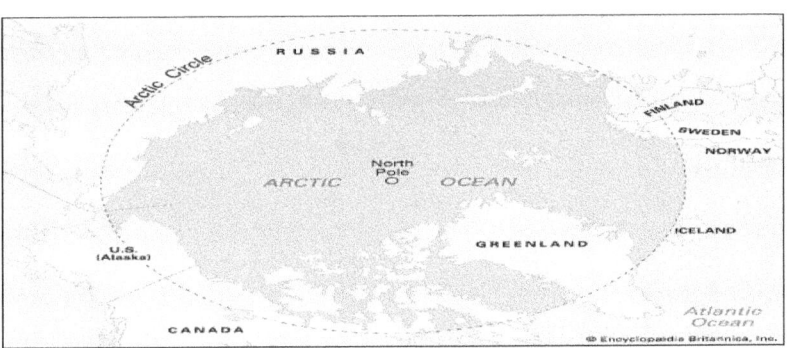

Image 2-9: Regions of the arctic circle area: (Sources: Encyclopedia Britannica, Mercopress October 20th, 2015)

- *Polar Bear*

It is the most well-known species in the Arctic ocean ecosystem. The polar bear is the largest carnivorous land mammal. Bears are still born on land, but they spend the majority of their lives on sea ice. When there is frozen in the Arctic Ocean, it can be found in and on edge of it. Alaska and its coastline support the entire polar bear population in North America, as well as several terrestrial locations. The polar bear survives thanks to its thick, dark-colored skin and bulky furry coat. It shields the Arctic Ocean from freezing temperatures. The polar bears are presently listed as a threat under the Endangered Species Act. Modern plans to allocate drilling in the Arctic Ocean could hurt the species even more. The polar bear's fur is made up of thousands of transparent and hollow hairs. Because of the reflection of light, the polar bear's hairs appear white.

Image 2-10: The polar bear in the arctic region

- *Greenland*

Greenland is the world's biggest island. It is known as 'Kalaallit nunaat,' which means 'People's Land.' Although Greenland is thought to have been inhabited since around 2500 BC, the province's raucous conditions kept it unknown to Europeans until the 10th century. The country is not very green; in fact, ice covers 80 percent of Greenland. Temperatures range from 0 to 10°C during the Arctic summer. During the winter, the ice sheet extends from North America to Russia. Greenland's weather is notoriously unpredictable, making travel there difficult. Greenland is home to over 4 million people who live in this winter wonderland. Among these are the Arctic indigenous peoples. Greenland is home to two ecoregions; Kalaallit Nunnat high arctic tundra and Kalaallit Nunnat low arctic tundra.

Table 2-5: List of Ports and Harbours at Arctic ocean

Region	Country/Territory	Region	Country/Territory
North America	Canada, Churchill	North America	United States, Prudhoe Bay
	Tuktoyaktuk		Utquiagvk
	Greenland, Nukk Port and Harbour		Kivilina
Europe	Iceland, Akureyri	Europe	Russia, Murmansk
	Russia, Arkhangelsk		Naryan-Mar
	Belomorsk		Severomorsk
	Dikson		Tiksi
	Dudinka		Pevek
	Kandalaksha		Vitino
	Igarka		Sabetta
	Norway, Hammerfest		Norway, Honningsvag
	Kirkenses		Vaardo

References:

- A. Balasubramanian, 2014. OCEANOGRAPHY: THE PACIFIC OCEAN.
- Anon., 1989. Loss of biological diversity: A global crisis requiring international solutions, (National Science Foundation, Washington DC).
- Bargagli, R. 2005. Antarctic ecosystems: environmental contamination, climate change, and human impact. Berlin: Springer, pg 395.
- Bengtson, J.L., Boveng P., Franzén U., Have P., Heide-Jorgensen M.P., Harkonen T.J. 1991. Antibodies to canine distemper virus in Antarctic Seals. Marine Mammal Science, pg. 85-87.
- Charv is p., operto S., konnecke l.K., recq M., hello Y., houdry F., lebellegard p., louat R. & sage F., 1993. - Structure profonde du domaine nord du plateau de Kerguelen (océan indien austral): résultats préliminaires de la campagne MD66/keobs. C. R. Acad. Sci. Paris, pg. 341-347
- Coffin M.F., frey F.A., wallace p.J. Et al. (24 authors), 1999.- Kerguelen plateau-Broken Ridge: a large igneous province. pg. 1135-1142.
- Coffin M.F., pringle M.S., duncan R.A, gladczenko t.p., storey M., mueller R.D. & chahagan l.A., 2002. -Kerguelen hotspot magma output since 130 Ma. Journal of Petrology. pg. 1121-1139.
- David Michel and Russell Sticklor 2012. Indian Ocean Rising: Maritime Security and Policy Challenges.
- Delpech G., grégoire M., o'reilly S., Cottin J.Y., MOINE B. & MICHON G., 2004. - Felspar from carbonate rich metasomatism in the oceanic mantle under Kerguelen islands (South indian ocean). Lithos, pg. 209-237.
- Harris, P.T., MacMillan-Lawler, M., Rupp, J., Baker, E.K. 2014. Geomorphology of the oceans. Marine Geology pg. 4-24.

- Kerry, K.R. and Riddle, M. 2009. Health of Antarctic Wildlife: An Introduction. In: Kerry, K.R. and Riddle, M. (eds.). Health of Antarctic Wildlife. Springer-Verlag Berlin Heidelberg. pg. 470.
- Martínez, M.L., Intralawan, A., Vázquez, G., Pérez-Maqueo, O., Sutton, P., Landgrave, R. 2007. The coasts of our world: Ecological, economic and social importance. Ecological Economics pg. 254-272.
- PDMIN 2005. Pacific disaster management information network (Indian Ocean earthquake & tsunami emergency update).
- Ray G C, 1991. Coastal zone biodiversity patterns, Bioscience, pg. 490-498.
- Recq M. & charv is A., 1986. - A seismic refraction survey in the Kerguelen isles, Southern indian ocean. Geophyscal Joural of the Royal Astronomical Society, pg. 164-182.
- Rykaczewski, Ryan R., and Checkley Jr., D.M. 2008. Influence of Ocean Winds on the Pelagic Ecosystem in Upwelling Regions, Proceedings of the National Academy of Sciences of the United States of America.
- S. Tabata. 1975. The general circulation of the pacific ocean and a brief account of the oceanographic structure of the north pacific ocean 1 part I - circulation and volume transports.
- Santora, J.A. and Veit, R.R. 2013. Spatio-temporal persistence of top predator hotspots near the Antarctic Peninsula. Marine Ecology Progress Series, pg. 287-304.
- Steinberg, D.K., Martinson, D.G. and Costa, D.P. 2012. Two decades of pelagic ecology of the western Antarctic Peninsula. Oceanography, pg. 56-67.
- Strugnell, J., Cherel, Y., Cooke, I.R., Gleadall, I.G., Hochberg, F.G., Ibáñez, C.M., Jorgensen, E., Laptikhovsky, V.V., Linse, K., Norman, M., Vecchione, M., Voight, J.R. and Allcock, A.L. 2011. The Southern Ocean: source and sink? Deep Sea Research II. Pg. 196-204.

- Strugnell, J.M., Rogers, A.D., Prodohl, P.A., Collins, M.A., and Allcock, A.L. 2008. The thermohaline expressway: the Southern Ocean as a centre of origin for deep-sea octopuses. Cladistics.
- Tittensor, D.P., Mora, C., Jetz, W., et al. (2010). Global patterns and predictors of marine biodiversity across taxa. *Nature* pg. 1098–1101.
- UNEP 2002. Regionally Based Assessment of Persistent Toxic Substances. Antarctica Regional Report, Global Evironoment Facitiy.
- Vecchione, M., Piatkowski, U., Allcock, A.L., Jorgensen, E. and Barratt, I. 2009. Persistent elevated abundance of octopods in an overfished Antarctic area. pg. 197-203.
- Weimerskirch, H., Inchausti, P., Guinet, C., Barbraud, C. 2003. Trends in bird and seal populations as indicators of a system shift in the Southern Ocean. Antarctic Science, pg. 249-256.
- White, M.G. 1984. Marine benthos. In: Laws, R.M. (ed.) Antarctic Ecology. vol. 2. Academic Press, London, pg. 421-461.
- Wulff, A., Iken, K., Quartino, M.L., Al-Handal, A., Wiencke, C., Clayton M.N. 2011. Biodiversity, biogeography and zonation of marine benthic micro- and macroalgae in the Arctic and Antarctic. (Capítulo 3, 23-52). In: Wiencke, C. (ed.) Biology of Polar Benthic Algae. De Gruyter, Berlin. pg. 23-52.

Websites:

- https://oceanservice.noaa.gov/facts/sevenseas.html
- https://kids.britannica.com/students/article/Pacific-Ocean/276242
- https://www.britannica.com/place/Atlantic-Ocean
- https://30a.com/facts-about-the-atlantic-ocean/
- https://zeenews.india.com/hindi/world/photo-gallery-know-some-interesting-facts-about-bermuda-triangle-mystery/796132/who-find-bermuda-triangle-796134

- https://www.softschools.com/facts/geography/atlantic_ocean_facts/1262/
- https://hindi.webdunia.com/interesting-and-exciting/interesting-facts-of-indian-ocean-121021900105_1.html
- https://timesofindia.indiatimes.com/world/explained-all-you-need-to-know-about-newly-names-southern-ocean/articleshow/83627992.cms
- https://www.worldatlas.com/articles/10-important-facts-you-must-remember-about-the-southern-ocean.html
- https://kids.britannica.com/students/article/Arctic-Ocean/272940
- https://www.livescience.com/15177-gallery-bermuda-triangle.html

Chapter 3
Marine Habitats

The habitat is an environment that renders complete set of life requisites to organismsin order to survive. More than 2,35,000 marine living organisms have been identified by marine biologists, and the number is growing as new ones are discovered. Depending on the temperature of the water, marine habitats can be extremely diverse. Tropical coral reefs are teeming with tiny, colourful fish, whereas polar waters have fewer species that have adapted to water that can be colder than the temperature at which water usually freezes at 0°C. Marine habitats are divided into two types: coastal habitat and open ocean habitat. In contrast, it is divided into two zones: pelagic and demersal.

Several organisms survive in various habitats such as beaches, mudflats, rocky shore, seagrass, coral reef, estuary, surface water, deep water, and so on. Coastal habitats can be found in an area that has expanded from the point where the tide comes in on the shoreline to the rim of the continental shelf. Oceans, estuaries, on-shore ecosystems, and coral reefs are all part of the marine ecosystem. In general, marine organisms live on the ocean's surface within the sunlight zone; enough sunlight can support photosynthesis by marine algae, which either directly or indirectly provide a source of food for the vast majority of marine organisms.

3.1 Rocky Shore

It is simply observed that vegetation and animals on tidal rocky shores tend to congregate in bands or zones close to the water's edge. A rocky shore is a section of seashore where solid rocks predominate. It occurs as a result of marine erosion of the overburden and bedrock due to a combination of sea-level rise and wave action in areas where there is low sediment supply. The environment and properties of rocky shores are primarily related to the mode of formation, with limestone, basalt, and granite being the most common.

Living organisms must be able to adapt to constantly changing conditions such as changes in water temperature, levels, and crashing waves, as well as avoid predators. Rocky pools have physical characteristics, such as unshaded shallow pools that show significant changes in temperature, salinity, oxygen, pH, and so on. It's a natural world with unique habitats. Rock pools have always piqued the interest of naturalists. Organisms that are particularly sensitive to desiccation can often extend their range up the shore in shaded, damp crevices.

The rocky shore may appear calm, but there are huge changes with the rising and falling of the tide twice a day. Intertidal species are tiny and sessile, and they are distributed across steep physical gradients that occur over short distances. Low tides regularly expose rocky intertidal habitats to above-ground conditions. The upper limit of the intertidal area is extremely high spring tides, while the lower limit is extremely low spring tides. Some organisms at the top of the shore may only be wetted twice a month, while others at the bottom of the shore may only be exposed briefly twice a month.

A rocky intertidal area has a rich environment that consist steep rocky cliffs, platforms, rock pools, and boulder fields. Erosion is a term used to describe the permanent action of tides and waves. Other factors such as breeze, light, temperature, oxygen, and physical factors contribute to a compound environment. The Rocky shores are important for a variety of reasons, including providing a nursery area for many molluscs, crustaceans, and fish.

They provide shelter in areas where seaweeds reduce wave action. Rare and threatened species rely heavily on algal beds for food. At low tide, the rocky shore provides food for wading birds. Like other intertidal habitats, rocky coasts are found at the land-sea boundary, so their physical and biological energy is influenced by both marine and terrestrial processes.

The rocky areas have open ecosystems that are dominated by vertical environmental inclines. In the WIO, the intertidal rocky shore is focused on varying degrees of commercial and recreational fishing survival. In rock pools and shallow intertidal lagoons, spearfishing is also common. It has also been reported that certain sponges, echinoderms, and molluscs are being collected for medical purposes. Organisms display unique supply patterns in response to a range of factors operating at different spatial scales. In the Western Indian Ocean region, intertidal rocky coasts have significant ecological, socioeconomic, and conservation value. The rocky shore has a wide range of niches for fauna to inhabit; rocky pools are a distinct habitat on the rocky shore. It is varies in size from a small depression to crevices. Submerged rocks that are never exposed are affected by environmental factors such as insolation, salinity, and temperature. Even at high tides, rocks that are never submerged receive the face of the rock near the sea as well as the full collision of the waves. The action of the waves does not dislodge rocks and loose stones along the shoreline. The stability of the rocks and stones is critical to their colonisation.

The rocky shore's solid bottom is frequently invaded by many sessile forms because it provides secure places for attachment.

Different fauna have different effects on the substratum. The ability to remain in a state of suspended activity for extended periods of time is a distinguishing feature of coastal animals. These are the aestivation or hibernation periods of terrestrial animals. The emergence of suspicious mechanisms in animals living on exposed rocks is a common occurrence.

3.2 *Sandy Shore*

It is also referred to as 'Beaches'. The sandy shores are frequently known to be devoid of life, made up only of sand, shells, and the occasional piece of driftwood. This vibrant habitat is home to a range of several species, each specially adapted for life over or beneath the sand. Tides are supplying food, oxygen, and nutrients; it carries left waste products. The sand is essentially rather complex. It is made up of abiotic things such as small, loose pieces of rock, soil, and minerals, as well as biotic things such as shells.

There is an inclination for the concentration of well-built particles of the sand in the upper part of the beach. Which thwarts the animals from in places whether severe wave action is. Sands are made up of quartz, calcium carbonate, or volcanic fragments. The plants and animals have adapted a wide array of adaptations. The sand and water are in stable motion; several creatures are proficient burrowers and burrow down into the sand to avoid being washed away by strong waves. These organisms are predated by aves and mammals.

It has been drawn to human activity since ancient times. It provides goods and services, such as forming as natural coastal cover, groundwater reserves for drinking water, coastal protection by providing fresh water, and nourishing economical activities such as fisheries and amusement. The world is currently being impacted by overpopulation, subsidence, and sea-level rise. They also supply numerous societal benefits, such as leisure opportunities; more information on human benefits derived from sandy shores are discussed in the section Ecosystem services. The newly formed element with sharp edges is not the proper substratum for burrowing animals because it makes penetration into the sandy substratum difficult.

The fine sand particles offer good scope for the inhabitation of some macro fauna as well as infauna called meiofauna or interstitial fauna. The beaches have a small permanent population. There are only a few seaweeds and seagrasses capable of rooting themselves. The majority of the flora and fauna are bottom dwellers. The segmented worms, crabs, molluscs, and a few echinoderms make up the vast majority of sand dwellers. Despite the lack of flora and a hiding place, the beaches are home to a diverse range of fauna. Burrowing has first and foremost provided protection against the pounding action of the waves as well as the force of diminishing waves. Because the majority of them live in burrows, they have developed certain characteristics in common, such as digging organisms, ciliary mode of feeding, and some mode respiratory devices. The majority of them feed on the decomposing organic matter in the sand. They swallow the sand containing this organic detritus and digest it in the same way that an earthworm would. To avoid the negative effects of decreased salinity, animals living near the surface will have to burrow to deeper levels. Various beach organisms, such as the burrowing prawn *Callianasa* and *Upogebia*, have only very narrow stenohaline. The water on the beach changes in temperature and salinity, and it also contains oxygen, which the organisms can use.

3.3 *Mud Flats*

The mudflats are mud, silt, and clay deposits. The term "mudflats" refers to land near a body of water that is frequently flooded by tides and is usually infertile. Mudflats are formed by the deposition of mud by tides or rivers. It is a vibrant habitat, and its survival is dependent on maintaining the equilibrium between the rate of sediment deposition from the water column and sediment erosion due to tidal and wave action. The mud is also known for its elevated organic carbon content and hence serves as a very good food source for many of the mud-dwelling organisms. This is common, to the detriment of other coastal habitats such as salt marshes. Because the majority of a mudflat's sediment area is within the intertidal zone. A large number of migratory coastal birds visit mudflats.

It acts as a barricade to waves from eroding land in the centre. The loss of these tidal flats will make coastal areas more vulnerable to the forces of erosion and also floods. According to a recent global remote sensing study, Australia, Brazil, Canada, China, India, Indonesia, Myanmar, and the United States account for roughly half of the international extent of tidal flats. These cover more than 44% of the Asia of a coastal boundary.

Did You Know...?

- Rocky pools can be a harsh place to live.
- The Colour of sand is variable such as white, tan, black, green, or even pink!
- Several human activities can impact the use of mudflats by feeding and roosting birds.
- Intertidal zones are transition lines between land and ocean.

Various development projects such as land claims and the building of hard coastal defences can lead to undeviating loss and the fragmentation of naturally mobile coastal habitats such as mudflats. The result of sea-level rise is that the low water mark migrates landward, while coastal ramparts prevent the high water spot from migrating landward. It causes a phenomenon known as "coastal squeeze" and a decline in habitat. The introduction of a novel or non-native species, such as cord grass, can have a significant impact on the mudflats ecology.

3.4 Intertidal Zone

It's also referred to as the seashore or the littoral zone. During low tide, the intertidal zone was exposed; however, during high tide, it was submerged. It is divided into four subzones: the spray zone, the high tide zone, the middle tide zone, and the lower tide zone. These four zones differ from one location to the next. Organisms that live in tidal pools have adapted to drastic changes in salinity. They can withstand fluctuating salinity levels. The level of salinity is rising as the temperature rises. It is the splash zone, which is the highest area between the terrestrial and marine realms.

The highest high tide barely covers the high intertidal zone. Byssal threads, Holdfasts, Muscular feet, and other devices are used by the organisms. During tidal cycles, the middle littoral zone is regularly exposed and submerged. This area is a mark between the high tide line and the low tide line. However, it is mostly exposed during low tide. It has the most diverse fauna and flora groups. During the lowest low tide, the lower zone is exposed. The tidal range is 0.4-6 m, depending on the geographical location of the intertidal zone. The area can be quite large at times, stretching for several kilometres. Many different types of living organisms can be found in rock pools, crevices, and tidal pools. They found it easily during the water explosion.

High tidal organisms get enough food in the form of detritus, seaweeds, and microorganisms. It provides many major ecosystems that are also important to human livelihood in various ways. The intertidal zone's diversity is dwindling. The loss of biodiversity is caused by a variety of factors. Increasing human population on the coast, as well as other factors such as excessive and damaging use of the intertidal area, over-human exploitation, natural disasters, and hand overfishing, are all to blame for threatening several intertidal areas.

Image3-1: (A)Highly exposed Rocky shore, Gir-Somnath, India

(B) Wave splash with the sandy shore at Devbhoomi- Dwarka, India,

(C) Loose mudflats in low tide, Koliyak,

(D) Exposed intertidal area in Gosa, India.

3.5 Mangrove

Duke (1992) defined mangrove as a tree, a shrub, a palm, and a fern, generally exceeding a one-half meter in height and which generally grows over mean sea level in the intertidal zone of marine coastal environments or estuarine margins. This definition is acceptable, with the exception of ground ferns, which should be considered mangrove associates rather than true mangroves. Mangroves found at the junction of terrestrial, estuarine, and near-shore marine ecosystems have enormous ecological and economic significance. Mangroves are the only species of trees in the world that can tolerate saltwater. The protection of mangroves is of extreme importance to sustain the health of the fragile environment. They are various kinds of trees up to medium height and shrubs. It can grow in saline coastal sediment habitats. Mangrove forest services are provided a value of at least $1.6 billion per year worldwide. It is vibrant and unique ecosystem is increasingly threatened and depleted. Species planted should be attuned to the salinity and substrate conditions.

This ecosystem can be found on the saltwater coasts of 118 tropical and subtropical countries. Mangrove roots provide nurturing environments for juveniles of thousands of fish species ranging in size from one inch to ten feet. According to a survey, 70% of true mangrove species are in two critically endangered categories and three endangered categories. Because of extensive habitat loss, at least 40% of the animal species restricted to mangroves are on the verge of extinction. Mangrove habitat loss has an impact on the local communities that rely on them, either directly or indirectly.

Mangroves provide firewood, wood products such as timber, poles, and posts, as well as non-wood products such as fodder, honey, wax, tannin, dye, and plant materials for thatching. Along with cyclones, mangrove wetlands and forests provide protection. It also prevents to keep the coast from eroding. They supply feeding, breeding, and nursery grounds for several economically vital fish, prawns, crabs, and molluscs. By exporting nutrients and detritus, they act as fine webs and help to develop the fishery production of nearby coastal waters. They provide habitats for wildlife ranging from migratory birds to estuarine crocodiles, tigers, and others. They are sites of accretion of sediment, and they act as 'sink' for carbon and nutrients. Mangrove forests filtering abilities protect critical coral reefs and sea grass beds from damaging siltation. It is an easy way of replication that is taking place in the mangrove forest.

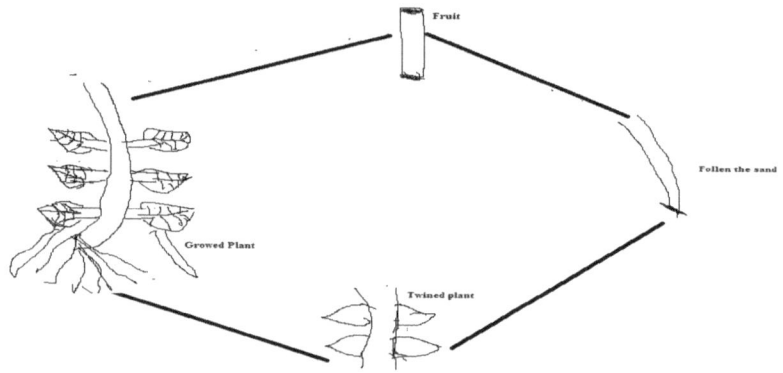

Image 3-2: Life cycle of Mangrove plants

Image 3-3: Incredible dense mangrove ecosystem at Pichavaram estuary, Tamil Nadu, India

3.6 Estuaries

The estuary is an integral point of the seashore. It is the tidal mouth of the sea and river. It is a semi-enclosed water body where fresh water and mix close interaction between land and sea. The wave height is extremely low. It is divided into three sections: marine or lower estuaries, which have a free connection to the open sea; middle estuaries, which lack strong salt and fresh water; and higher and fluvial estuaries, which are subject to daily tidal action. The sea has a salinity of about 35 ppt, whereas freshwater has less than 0.5 ppt.

Estuaries are classified into several types; four types of estuaries have been identified based on geomorphology. i.e., drowned river valleys, bar built, fjords type, and tectonic estuary. Typically, an estuary has a gradation of salinity from the mouth, where the water is completely saline, to the head, where the water is fresh. With an increasing flood tide, the freshwater flowing out to sea in most estuaries is insufficient to prevent an inflow of seawater up the estuary. The effect of Coriolis forces on opposing banks in any estuary wider than 400 m can result in a wide range of salinities. Estuaries are primary deposition areas, with an annual enlargement of 2 mm. Estuaries have long been used by humans for shipping, commerce, and more recently, recreation. The availability of shelter makes estuaries primarily safe as harbours, and their use by shipping inevitably leads to the advancement of docks and industry. Estuaries are the most productive natural ecosystems. Because of tidal action and current, the oxygen content of estuary water is higher than that of other bodies of water.

Estuaries also collect organic detritus from wetlands located throughout the estuary system. The majority of tropical estuaries are surrounded by mangrove forest–produced organic detritus, also known as a nutrient trap because it receives abundant nutrients from freshwater release. Some parameters in estuaries vary greatly, creating a stressful environment for organisms. Except for fjords, the water temperature in estuaries varies significantly due to their shallow depths and large surface area. It heats up and cools down extremely quickly in atmospheric conditions, which is due to freshwater input. Surface water has the highest temperature range, while deeper water has a lower temperature range.

The water in estuaries is murky due to the large number of substances present. The turbidity is the lowest near the mouth and increases with distance inland. The main ecological effect of turbidity is a noticeable decrease in light incursion. As a result, photosynthesis by phytoplankton and benthic plants declines, lowering productivity. Estuaries are valuable laboratories for scientists and students studying biology, geology, chemistry, physics, history, and social issues because they are transition zones between land and water.

3.7 Sea grass

Sea grasses are underwater flowering marine plants that play an important link in the ecosystem. It can be found at shallow depths in estuaries, bays, and lagoons with clean water. The development of sea grasses requires good water. The development of sea grasses necessitates clear water. It requires enough sunlight to penetrate the water to perform photosynthesis. The depth range of the meadow is determined by the availability of sunlight on the seafloor. Sea grass can grow at deaths of up to 32 meters depth. They can grow at depths of up to 65 meters in clear water. It grows in dense underwater meadows, some of which can be seen from space. About 100 million years ago, sea grass evolved. Seventy-two different sea grass species are classified into four major groups. There were four kinds of sea grass discovered: Zosteraceae, Hydrocharitaceae, Posidoniaceae, and Cymodoceaceae.

It can be found in shallow, salty, and brackish waters all over the world, from the tropics to the Arctic Circle. It gets its name from the fact that most species have long, grass-like leaves. Sea grasses are frequently confused with seaweeds, but they are more closely related to the flowering plants that resemble landing. Sea grasses are made up of three parts: roots, stems, and leaves. Flowers and seeds are produced by sea grasses. The leaves provide anchorage for other plants and have a symbiotic relationship with them. Many marine animals, including dugong, sea urchins, turtles, and some fish, are directly dependent on sea grass for food.

3.8 *Kelp Forest*

The kelp forest is the largest and most structurally diverse Phaeophyceae, encompassing a diverse group of genera known as the order Laminariales. Kelps are the primary constituents of the Atlantic and Pacific lower intertidal and subtidal zones. Globally, kelp forests support vast populations of fish and serve as a source of raw materials for the alginate industry. It lives in nutrient-rich, cold, shallow water.

Kelp forests use the sun for food, so they require being adequate to get the sun. Kelp can be used for several things such as the making of glass and soaps. They are also natural invasive species that we have introduced into the environment and that have an impact on them. The canopy, understory, and forest floor are the three layers of kelp forest. Every year, approximately 100,000 tonnes of kelp are harvested off the coast of California. It grows kelp for a significant business in California. These businesses employ a large number of people and contribute millions of dollars to the California economy. Kelps are approximately 3 to 10 times more dynamic than shallow-water phytoplankton and approximately 50 times more productive than oceanic phytoplankton. Few animals eat the living fronds of kelps as they decompose; chunks, particles, and organic molecules feed suspension feeders and deposit feeders.

Kelps are classified into three types: blades, stipes, and holdfasts. The blades are similar to the leaves of ground plants. They are the kelp's photosynthetic factories. It is here that light energy and water nutrients are combined to create food. Kelp stipes are resemble plant stems, but they do not perform nutrient movement functions like plant stems. Their primary function is to support the kelp blades. The holdfasts, which resemble the roots of land plants, aid in anchoring the kelp to hard surfaces. They are intricate ecosystems that support multiple interconnected food webs. They provide a diverse range of habitats for animals of all trophic levels.

Some abiotic factors influence kelp forests, such as light availability, which limits kelp depths. It requires only 1% of external irradiance to grow, it can grow deeper in clearer water. Temperature and nutrients both are correlated: as the temperature rises, the level of nutrients decreases. El Nino conditions, warm surface waters, depress thermocline, kelp growth declines, and kelps die, and nitrate is thought to be a key limiting nutrient. Water motion, massive waves rip kelp forests apart; some species are better adapted to dealing with high water motion than others. Hard substrate kelps, on the other hand, require a hard substrate to attach to (rocks); they cannot settle and grow on sand or other soft substrates.

3.9 *Coral Reef*

Corals are produced by small, polyps and cnidarians fauna. All polyps are zooids, colonies of similar species release gametes together. Polyps are all zooids, and colonies of similar species release gametes at the same time. As a result, not all corals can build reefs, and not all reefs are formed by corals; some modern reefs are formed by other organisms such as oysters, worm tubes, red algae, and cyanobacteria. Corals that do not form reefs can grow in the deep sea, temperate zones, and tropics. Coral reefs are normally restricted to the tropical and subtropical region, where the sea surface temperature averages at least 20°C. Coral reefs are commonly developed on the eastern boundaries of continents. Coral reefs normally thrive only in seawater with a constant salinity.

Coral reefs are classified into two types: shelf reefs that form along continental margins and oceanic reefs that form near islands. There are three types of oceanic reefs: fringing reefs, barrier reefs, and atolls. The red sea is longest known fringing reef stretches for about 400 kilometres. Barrier reefs are found offshore, separated from the shoreline by a lagoon. Australia's great barrier reef is by far the largest single biological aspect on the planet. Atolls are typically ring-shaped reefs with a few small islands protruding above the water's surface. The Marshall Island Kwajale in Atoll is the largest. This lagoon is 100 kilometres long and 55 metres deep.

Did You Know...?

- Mangroves are among the most productive ecosystems.
- Sea grass can absorb carbon up to 35 percent faster than Amazon rainforest.
- The kelp forests are seldom found much deeper than 100 to 125 feet deep.
- Coral reefs aid for better water quality.

Image 3-4: (A) Estuary(Source; Courtesy of The Wetlands Conservancy)

(B) Sea Grass (© Dr.DevAdhvan)

Marine Habitats

(C) Kelp Forest (Source;Enric Sala, National Geographic)

(D) Coral Reef (© Dr.Devanshi Joshi)

3.10 *Twilight Zone*

The disphotic zone is another name for the Twilight Zone. It is the layer immediately beneath the euphotic zone. There is enough light penetration in this zone for visual predators to see and even for a little photosynthesis. However, in the disphotic zone, the rate of respiration exceeds the rate of photosynthesis. With depth, the amount of light decreases. As a result, there is a scarcity of food. The photic zone is home to the majority of the ocean's organisms. The twilight zone can extend as deep as 800 metres in clear water mass. The photic zone is made up of the euphotic and disphotic zones. During the day, the centre layer of the world's oceans received only filtered sunlight.

This is due to the fact that seawater absorbs light. The depth of this zone is determined by the clarity or darkness of the water; in murky water, it can begin at only 50 feet. The disphotic zone reaches a depth of 3,300 feet. This zone typically ranges from 660 to 3,300 feet. Animals that live in the twilight zone have adapted to living in near darkness, cold water, and high pressure. Many of the animals in this zone have huge eyes, helping them see in the near dark waters. They are minute, murky, and slender. Several animals have large teeth and jaws. Various bioluminescent animals have distinct organs that generate light through a chemical reaction; other bioluminescent animals have glowing bacteria that live on them.

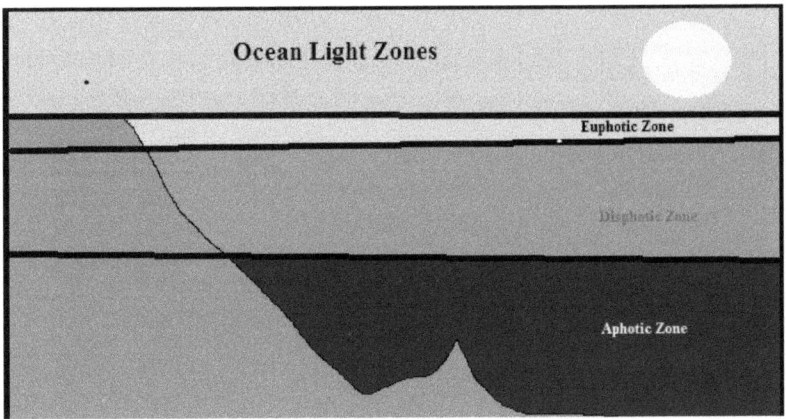

Image 3-5: Layers of light penetration

3.11 *Hadal Zone*

The hadal zone is also known as the aphotic zone or hadopelagic zone. It is also known as the ocean's deepest region. The term "hadal" was first proposed in 1956 by Anton Frederik Bruun. The hadal zone begins where less than one percent of sunlight can reach. As a result, the aphotic zone is very murky. The light produced by bioluminescent organisms is extra abundant than sunbeams. The production and emission of light by a living organism is known as bioluminescence. Hadal zones are not restricted to the ocean. Several lakes are also deep enough to support an aphotic zone. The hadal zone is found at depths ranging from 6,000 to 11,000 metres and is comprised of extensive but slight topographic V-shaped depressions. The deepest trenches are thought to be the least explored and most magnificent marine habitat. The hadal zone is characterised by harsh conditions such as low temperatures, nutrient deficiency, and extremely high hydrostatic pressures. Superior layers, fine sediment drift, and landslides are the primary sources of nutrients and carbon. The majority of organisms are scavengers and detrivores.

> **Did You Know...?**
>
> - The twilight zone appears deep blue to black.
> - Previously, the hadal zone was not known as distinct from the abyssal zone although the deepest divisions were sometimes called "ultra-abyssal".

Baikal Lake in Russia is more than 1600 m deep and has a large hadal zone. The increasing area occupied by the forty-six different hadal habitats covers less than 0.25% of the world's seafloor, yet trenches account for over 40% of the ocean's depth range. The Pacific Ocean contains the majority of the habitat.

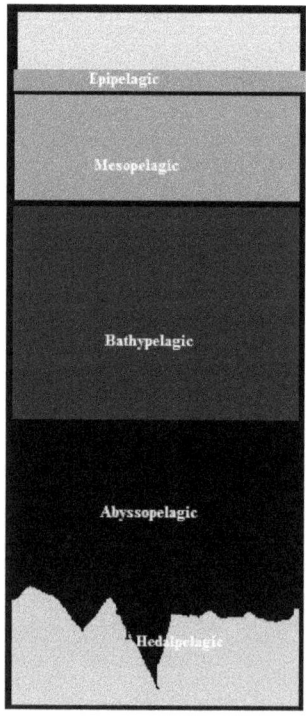

Image 3-6: Hadal zonation

References:

- A. N. Barkley, M. Gollock, M. Samoilys, F. Llewellyn, M. Shivji, B. Wetherbee, N. E.
- Hussey. 2019"Complex transboundary movements of marine megafauna in the Western Indian Ocean", Animal Conservation.
- Barbier EB, Hacker SD, Kennedy C, Koch EW, Stier AC, Silliman BR. 2011. The value of estuarine and coastal ecosystem services. Ecological Monographs 81:pg. 169–193
- Branch, G. M., Eekhout, S. and Bosman, A. L. 1990. Shortterm effects of the 1988 Orange River floods on the intertidal rocky-shore communities of the open coast. Transactions of the Royal Societyof South Africa 47, pg. 331-354
- Brosnan, D.M. and Crumrine, L.L. 1994. Effects of human trampling on marine rocky shore communities. Journal of Experimental Marine Biology and Ecology177, pg. 79-97.
- Crossland CJ, Kremer HH, Lindeboom HJ, Marshall Crossland JI, Le Tissier MDA.2005. Coastal Fluxes in the Anthropocene. The Land- Ocean Interactions in the Coastal Zone Project of the International Geosphere Biosphere Programme Series: Global Change - The IGBP Series, pg. 232
- Kelly, E. (ed.) 2005. The role of kelp in the marine environment. Irish WildlifeManuals, No. 17. National Parks and Wildlife Service, Department of Environment, Heritage and Local Government, Dublin, Ireland.
- Konar B, Iken K. 2018. The use of unmanned aerial vehicle imagery in intertidal monitoring. Deep-Sea Research Part II 147:pg. 79–86
- Lotze HK, Lenihan HS, Bourque BJ, Bradbury R, Cooke RG, Kay MC 2006. Depletion, Degradation, and Recovery Potential of Estuaries and Coastal Seas. Science 312: pg. 1806-1809
- Masselink G, Russell P, Rennie A, Brooks S, Spencer T. 2020. Impacts of climate change on coastal geomorphology and

- coastal erosion relevant to the coastal and marine environment around the UK. MCCIP Science Review 2020:pg. 158–189
- Menge, B.A. and Sutherland, J.P. 1987. Community regulation:variation in disturbance, competition, and predation in relation to environmental stress and recruitment. The American Naturalist130, pg. 730-757
- Menge, B.A., Lubchenco, J., Ashkenas, L.R. and Ramsey, F.L. 1986. Experimental separation of effects of consumers on sessile prey on a rocky shore on the Bay of Panama: direct and indirect consequences of food web complexity Journal of Experimental Marine Biology and Ecology 100, pg. 225-269
- Msangameno, D.J. 2013. Patterns and processes of benthic biological communities on the intertidal rocky shores of Unguja Island, Tanzania. PhD Thesis. University of Dares salaam
- Raffaelli, D. and Hawkins. S. 1996. Intertidal Ecology Chapman and Hall, London
- Ruwa, R.K. 1996. Intertidal wetlands. In East African Ecosystems and Conservation (eds. T.R. McClanahan and T.P. Young). Oxford University Press.pg. 101-127.
- Schiel, D.R. and Taylor, D.I. 1999. Effects of trampling on a rocky intertidal algal assemblage in southern New Zealand. Journal of Experimental Marine Biology and Ecology 235, pg. 213-235
- Scrosati, R.A., Knox, A.S., Valdivia, N. and Molis, M. 2011. Species richness and diversity across rocky intertidal elevation gradients in Helgoland: testing predictions from an environmental stress model. Helgoland Marine Research 65(2), pg. 91-102
- SubrataTrivedi, Abdulhadi A. Aloufi, Abid A. Ansari, Sankar K. Ghosh 2016."Role of DNA barcoding in marine biodiversity assessment and conservation: An update", Saudi Journal of Biological Sciences.
- Terlizzi, A., Fraschetti, S., Guidetti, P. and Boero, F. 2002. the effect of sewage discharge on shallow hard substrate sessile assemblages. Marine Pollution Bulletin44, pg. 544-550

- Thompson, R.C., Crowe, T.P. and Hawkins, S.J. 2002. Rocky intertidal communities: past environmental changes, present status and predictions for the next 25 years. Environmental Conservation, pg. 168-191
- Troell, M., Robertson-Andersson, D. Anderson, R.J., Bolton, J.J. Maneveldt, G., C. Halling, C, and Probyn, T. 2006. Abalone farming in South Africa: An overview with perspectives on kelp resources, abalone feed, potential for on-farm seaweed production and socio-economic importance. Aquaculture 257 (1-4), pg. 266-281
- Underwood, A.J. 2000. Experimental ecology of rocky intertidal habitats: what are what are we learning? Journal of Experimental Marine Biology and Ecology250, pg. 51-76
- UNEP/Nairobi Convention Secretariat 2009. Transboundary Diagnostic Analysis of Land-based Sources and Activities Affecting the Western Indian Ocean Coastal and Marine Environment. UNEP Nairobi, Kenya
- Wafar, M., Venkataraman, K., Ingole, B., Ajmal Khan, S., LokaBharathi, P. 2011. State of knowledge of coastal and marine biodiversity of Indian Ocean countries.*PLoS ONE* 6(1), e14613
- Worm, B. and Lotze, H.K. 2006. Effects of eutrophication, grazing and algal blooms on rocky shores. *Limnology Oceanography*51(1), pg. 569-579
- Yorath, C.J. and Nasmith, H.W. 2001. The Geology of Southern Vancouver Island; a field guide. Orca Book Publishers

Websites:

- https://encounteredu.com/cpd/subject-updates/learn-more-what-are-marine-habitats
- https://courses.lumenlearning.com/boundless-microbiology/chapter/aquatic-microbiology/
- http://www.solentforum.org/publications/key_publications/habitat_info_pack/mudflats.pdf

- https://www.ecoshape.org/en/landscapes/sandycoasts/sandy-shore-environments/
- https://oceanservice.noaa.gov/facts/intertidal-zone.html, http://blog1.miami.edu/sharklab/wp-content/uploads/sites/28/2018/07/MODULE-1-Ocean-and-Coastal-Habitat-SECTION-4-Intertidal-Zones.pdf
- https://www.opb.org/news/article/ estuaries-disappear-american-west-coast/
- https://landsat.gsfc.nasa.gov/article/kelp-forests-end-earth
- https://wildcoast.org/
- www.marine.usf.edu

CHAPTER 4
MARINE LIFE

The ocean has yet to be fully explored, and reaching the bottom is difficult due to its complexity. The term "marine life" refers to the living organisms that surround marine life. Flora, fauna, and microorganisms are all types of organisms. Oceans would be a more appropriate name for this blue planet. The ocean has long been a mystery, although its interior is largely accessible. Even if it does not capture the sea monsters that sailors once imagined, it still raises several questions for many scientists. Researchers discovered that due to the difficulty of safely reaching the bottom, less than 10% of the deep ocean is explored. It has only explored a small area of the deep ocean floor. The investigators discover that animals have a wider range of morphological variations than terrestrial animals.

Deep water life also exhibit different set of challenges. There is no sunlight and thus no photosynthesis process a few meters deeper. Deep-sea communities and the middle layer of water must rely on photosynthesizers found in the sunlit exterior waters. The deep-sea fauna is sustained by minuscule phytoplankton, zooplankton, and decomposing particles as they submerge. There isn't much plant life in the ocean because the concentrations of critical nutrients are lower than on land. The temperature stratification of the ocean determines the universal provider of the few nutrients that exist. Surface waters in tropical areas are always mild, whereas superior waters in temperate regions are warm in the summer and freeze the rest of the year. The thermocline is a thin zone that exists beneath the well-mixed exterior layer. It is impossible to overestimate the importance of frozen and heavy water in the food chain.

The difference between tenderness on the outside and cold on the inside is so great in the tropical zone that even hurricanes and typhoons cannot combine the two. As a result, nutrients and phytoplankton that rely on them are depleted in tropical waters. Because tropical seas are devoid of these clouds of tiny life, they are usually crystal clear. In temperate zones, winter storms can agitate the ocean, bringing some nutrients to the surface. The most common multicellular organisms on the planet are animals. In distinguishing photosynthetic organisms such as seaweed and flora, as the animals are unable to produce their food, they must rely on the food of others. The various types of marine fauna follow the traditional way of classifying them into two major groups: vertebrates and invertebrates.

The marine protists are a minute size that is eukaryotic and heterotrophs in nature. Many scientists believe that high protist diversity exists in oceans, deep-sea vents, and river sediments, implying that vast numbers of eukaryotic microbial communities have yet to be discovered. Furthermore, some modern authors argue that should be excluded from the traditional definition of protists, limiting protists to unicellular organisms. Protists are classified into four types: algae, protozoans, slime molds-slime nets, and other mixotrophs.

> **Did You Know...?**
>
> - It is approximated that at least 97% of all groups of fauna are invertebrates.
> - All key groups of invertebrates have marine representatives and many are exclusively marine.
> - Only some groups have successfully invaded terrestrial life

They are a great and varied group. Some microalgae and diatoms are single-celled however some are multicellular. exa. Red algae, Brown algae, Diatoms, etc. Protists that resemble animals are referred to as protozoa. Zooflagellates, foraminifera, radiolarians, and some dinoflagellates are examples of marine protozoans. It can able to be seen in a low-power microscope as well. Mainly, protozoa consist of solitary cells and heterotrophs. Protozoa can move. The protozoa engulf, feed on, and digest other organisms. Some are herbivores, while others are predators. The two most common types of fungus-like organisms are slime molds and water molds. Slime molds have motile cells at various stages of their life cycle. It moves slowly as it searches for the decomposing matter to eat. When food is scarce, entity cells swarm together to form a blob-like mass resembling 'dog vomit.' Water molds are commonly found in wet soil and on the surface of bodies of water. Many plants are pathogens that cause crop damage. It is capable of contaminating plants such as grapes, lettuce, and potatoes.

Porifera, or pore bearers, are thought to be related to the first multicellular animals, which were likely simple colonies in which a few cells specialized for roles such as feeding and protection. Pinacocytes and a few tube-like porocytes cover the surface, and a minuscule canal allows water to pass through. Sponges are suspension feeders, which consume food particles suspended in water. Sponge feeders, also known as filter feeders, are a type of suspension feeder that vigorously filters the food particles. In general, the choanocytes are located in an extra compound arrangement in marine sponges.

Sponge species can be found from the poles to the tropics, but the vast majority of them live in shallow tropical waters. Sponges can grow into a variety of shapes, including branching, tubular, round, or volcano-like masses that can grow to be quite large. Encrusting sponges are thin, brightly colored growths that grow on rocks or dead coral. The Venus flower basket sponge has a grass-like skeleton of fused siliceous spicules and lives in deep water sediments. Cliona excavates slim channels through calcium carbonates like oyster shells and corals. Several marine porifera species produce chemicals that may be commercially important to humans.

Cnidarians are mostly marine creatures that live in water. Cnidaria is composed of 9000 different species. The majority of cnidarians live in tropical and neotropical seas. The body is radially symmetric, diploblastic, and devoid of a coelom. The tentacles are surrounded by a large cavity known as the gastrovascular cavity or coelenterons. Cnidarians are mostly marine animals like coral, hydras, jellyfish, portuguese men of war, sea anemones, and sea pens. Cnidarians have two life stages: polyps and medusae, as well as an embryonic or larval form called a planula. Digestion takes place both intracellularly and extracellularly. It has a mouth-shaped opening, known as a stomodeum. Food particles are extracellularly digested in the coelenteron before being distributed to the body wall for intracellular digestion. They primarily capture and digest much larger prey than filter feeders such as sponges. The primary function of nematocysts is to capture prey. Some nematocysts contain toxic materials. After ingestion, food sources pass into the gut and are digested. To complete the respiration and excretion systems, simple diffusion is used. On Australia's northern coast, the Great Barrier Reef runs parallel to the shoreline and frequently surrounds a lagoon, shielding landmasses from high surf. The best of the reef lagoon ecosystem is dependent on the coral's endosymbiotic algae, and corals stop growing when their algae are discarded due to global warming and possibly pollution.

> **Did You Know...?**
>
> - All cnidarians are carnivores.
> - Approximately 9,000 species of sponges are marine dwellers.
> - Some worms can survive in enormously deep trenches, such as the Pacific Ocean off Galapagos Islands

Because of the toxic substances found in cnidarian nematocysts, all cnidarians are likely to have an impact on human physiology. Most are not harmful to the human body, but some can impart a painful sting such as Physalia (Portuguese man o' war) and Actinodendron. It is usually harmless cnidarians species, but it can be noxious quantities or to a sensitive person. The only cnidarians that are generally fatal to humans are the cubomedusae. Heart-stimulant, antitumor, and anti-inflammatory properties are found in extracts of various cnidarians, primarily anthozoans. The fundamental body plan is structurally simple and very thriving. The epidermis is on the outside, whereas the gastrodermis is on the inside.

Marine worms can survive in a marine environment. Many of these worms have tentacles that exchange oxygen and carbon dioxide. Tube worms include some marine worms, such as the giant tube worm. It can withstand temperatures of up to 90°C and can survive in the vicinity of underwater volcanic activity. In recent years, marine worms have been observed ingesting microplastic particles found in the oceans. Many scientists are concerned about this trend because marine worms are a vital source of food for many fish and wading birds.

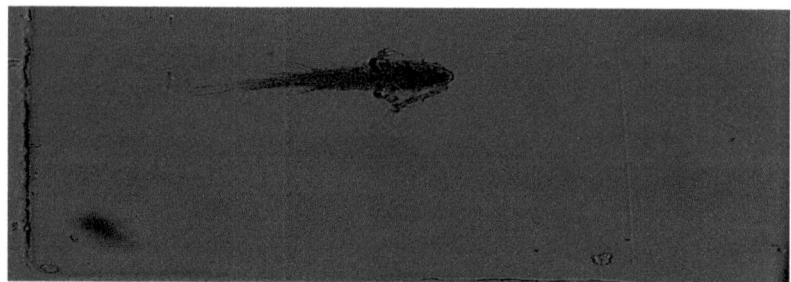

Image 4-1 (A) Marine Protozoa

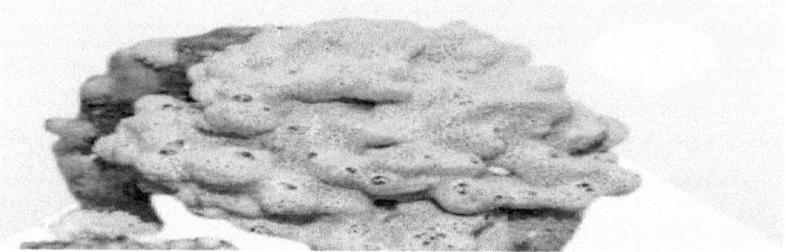

(B) Porifera

(C) Cnidaria

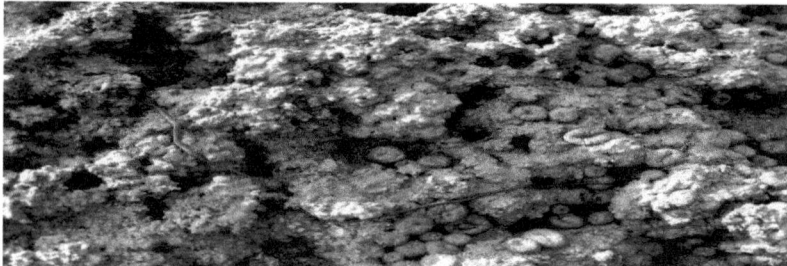

(D) Marine worms

With a density of up to 20 million individuals per square meter, marine nematodes are the most abundant metazoans in marine sediments. It is an estimate of global nematode species diversity that has changed significantly over the last 15 years, but there is growing agreement that there are around one million species. It is well understood how important they are as bioindicators in marine ecology. The study of nematode communities suggests a variety of advantages for assessing the quality of freshwater, marine, and terrestrial ecosystems. These organisms are diverse and abundant, making sampling easier; they belong to various trophic groups and are closely related to their environment.

Marine Platyhelminthes are typically brightly coloured creatures. The bright species are easy to spot while diving. On both sides, they are soft, unsegmented, and symmetrical. They have a very flexible body and no skeleton. They are deficient in both the circulatory and respiratory systems. There is no fossil record of Platyhelminths. Some Platyhelminthes trace fossils have been reported, and fossil trematode eggs have been discovered in Egyptian mummies and Pleistocene period ground sloth dung.

Arthropoda makes up roughly 75% of all animals on the planet. It has the most species as well as the most people. It is composed of three components: segmented bodies, jointed legs, and a hard exoskeleton. Nearly a million arthropoda species exist, with more than 90% of them being insects, leaving less than 10%, or 85,000 species, with only three major marine groups. The most well-known Crustacean is the marine Crustacean. It is home to over 30,000 different species. The entire marine class Pycnogonida, also known as "sea spiders" (500 species), and the entire marine class Merostomata, also known as horseshoe crabs are some of the examples. The three major groups of marine arthropods are horseshoe crabs, sea spiders, and crustaceans. Crustaceans include barnacles, shrimp, lobster, mole crabs, and true crabs. Crustaceans, among other things, can shed their shells, metamorphose, and regrow lost limbs. During the roughly 250 million years since arthropods were first appeared in the ocean, their basic body has adapted to an enormous variety of niches in which they can live, resulting in the emergence of new species.

The Molluscs are the most diverse marine phylum; it encompasses approximately 23% of all marine diversity. In terms of evolution, molluscs have been around for a long time. According to fossil records, the first mollusc was discovered around 500 million years ago during the Cambrian period. The planet is home to approximately 85,000 different species. The most important class is Gastropoda, which contains more than 80% of all living molluscs species. This class includes species from the aquatic, freshwater, and terrestrial environments. Molluscs can be found from the intertidal zone to the benthic zone. Molluscs are very plentiful and form a key link in the food chains. Molluscs are distinguished by their uniqueness in that they are soft-bodied, non-metameric, triploblastic, and coelomate. Invertebrate with a bilaterally symmetrical head, a ventral muscular foot, and a visceral mass protected by an external calcareous shell and covered in a thin, fleshy mantle. Molluscs have a massive diversity of structure and habit. It ranges from the supratidal zone to deep-sea hydrothermal vents. The importance of gastropods and bivalves in the economy cannot be overstated.

Did You Know...?

- The largest phylum of animal on Earth devoid of doubt is Arthropoda.
- Molluscs are the second-largest phylum of invertebrates.
- The echinoderms are exclusively marine dwellers.

There is a wide range of size variations observed. Molluscs include snails, slugs, mussels, clams, oysters, octopuses, squids, etc. The beaches have a lower diversity of burrowing forms. Molluscs are extremely successful in the oceans, and as a result, they are the most abundant invertebrate.

The echinoderm is a spiny and skinny animal. Marine organisms are those found on the seafloor or near the coast. Although some species are found in the Arctic, they can be found in all marine habitats. Endoderms, mesoderms, and ectoderms are organized in three layers of the body, which are represents radial. It is important in ecology for food sources, predators, nutrient recycling, and so on. Echinoderms can vanish or damage body parts. It feeds in a variety of ways, with some species consuming animal remains on the ocean floor. Others do not filter plankton through their mouth pores.

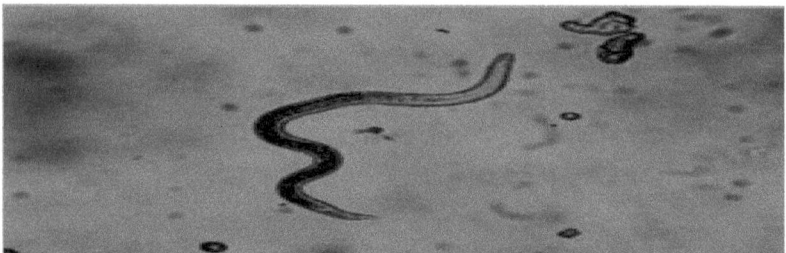

Image 4-2 (A) Marine Platyhelminthes

(B) Crustacean

(C) Molluscs

(D) Echinoderms

The marine reptiles evolved from prehistoric terrestrial reptiles that eventually colonized the sea. True marine species include sea turtles, marine iguanas, and sea snakes. The marine reptiles were forced to adapt to a salty aquatic environment and develop diving-adapted breathing abilities. The reptile's streamlined bodies are similar to those of fish. Their dry skin is covered with scale to prevent water loss. To keep the eggs of marine reptiles from drying out, they have a leathery shell. Reptiles are cold-blooded animals that are poikilotherms and ectotherms. Since various reptiles were the primary marine predators between 180 and 90 million years ago. Reptiles evolved to live on land and were not well suited to aquatic life. The most numerous group of marine reptiles is the sea snakes. It can be found in the Indian and Pacific Oceans, from the east coast of Africa to the Gulf of Panama, in tropical and subtropical waters.

Sea birds are also known as marine birds. Seabirds have evolved to be able to live in the sea. Through includes only about 3% of the estimated 9,700 bird species. It is found all over the world, from the North Pole to the South Pole, and has a significant impact on marine living organisms. Birds are homeotherms, or "warm-blooded". They are endothermic organisms. Sea birds are birds that spend the majority portion of their lives at sea and feed on marine organisms. The marine birds live longer and breed later. Most sea birds build nests on land, breed in large colonies, mate as enduring pairs, and care for their offspring. It is descended from numerous groups of terrestrial birds.

> **Did You Know...?**
>
> - The numbers of true marine species signify only 1% of all the reptile species that survive today.
> - Marine birds are spending a noteworthy part of their lives at sea and feed on marine organisms.
> - The species richness of marine mammals is peaked at around 40° latitude equally north and south.

Sea birds have an amazing craving, they require a lot of food resources to supply the energy required to maintain their temperatures stable. Among the seabirds are many species of wedding shorebirds with webbed feet. Because they do not swim frequently, they are not technically sea birds. Marine birds are also famous for their capacity to travel enormousdistances: Arctic terns *(Sterna paradisea)* has the longest recorded voyage of any animal. It can fly more than 80,000 km from Arctic breeding grounds to the Southern Ocean and back each year.

Marine mammals are widely distributed throughout the earth, but their distribution is irregular and coincides with the productivity of the oceans. Marine mammals have undergone foremost adaptations that allow them to live in an aquatic habitat. Other adaptations of marine mammals include a slower heartbeat during dives, reduced blood flow to non-vital organs, a bizarrely high hemoglobin count in the blood, and an unusually high myoglobin count in the muscles area. Most marine mammals have varieties that are equal to or smaller than one-fifth the size of the Indian Ocean. Marine mammals include herbivores, filter feeders, and top predators. The marine mammal group evolved from a diverse group of terrestrial mammals, each of whose ancestors ventured back into the marine environment independently. Several mammals evolved in an ocean environment with streamlined bodies and paddle-like limbs. Marine mammals frequently migrate seasonally.

Marine mammals are adaptations to a marine lifestyle that differ greatly between species. The most significant structural modification to the bodies of cetaceans (porpoises, dolphins, and whales), sirenians (manatees and dugongs), and pinnipeds is the loss of hind limbs (seals, sea lions, and walruses). Because of the rotation of the earth, primary surface currents flow clockwise in the Northern Hemisphere and counterclockwise in the Southern Hemisphere. This has several implications for marine life on both the east and west sides of ocean basins. Cetaceans and sirenians are entirely aquatic, whereas seals and sea lions are only partially aquatic; they spend the majority of their time in the water but must return to land for essential activities such as mating, breeding, and molting.

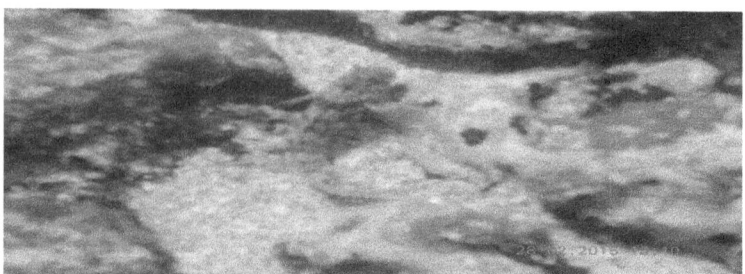

Image 4-3 (A) Marine Fish

(B) Reptile

(C) Sea Bird

(D) Mammal

The plants have established themselves in the oceans. The most common type of marine vegetation found in a marine environment is seaweed, which can be found from the pelagic to the bathypelagic zones. Many different types of vegetation, mostly in tiny forms, can be found in the ocean's upper layers. The solar rays which penetrate the water, as well as the ability of seaweed to capture rays of a specific wavelength at different depths, determine the distribution of vegetation. Seaweed is available in a wide range of sizes, from microscopic to massive. Some seaweeds have thallus that is not segmented and can be thread like, ribbon like, plate like, or leaf like in shape. Grasses from the Potamogetonaceae and Hydrocharitaceae families dominate the higher plant life of seas and oceans. Based on the geographic distribution of benthic seaweeds and seagrasses, the world's oceans are divided into five regions: Arctic and Antarctic (two polar zones) with slightly more vegetation in the sublittoral, Boreal and Austral (two temperate zones) with abundant growths of Phaeophyceae and seagrasses (littoral and sublittoral), and numerous species of Chlorophyceae and Phaeophyceae (one tropical zone).

Marine algae are an assemblage of plants that are in either saltwater or brackish water. It can be found between high tide and low tide on the coast, as well as in the subtidal zone up to a depth of 0.01 percent photosynthesis sunlight. The size varies from a minute unicellular variety of microns to large macroscopic multicellular forms up to 70 meters extended and growing at up to 50 cm per day. It beautifies the underwater landscape while also providing direct value to man as a food and industrial raw material.

> **Did You Know...?**
>
> - The Algae is found in the coastal area between high tide to low tide and subtidal area up to a depth where 0.01% photosynthesis sunlight is obtainable.
> - The mangrove species are about 80 different types.
> - About 40 times more animals occur in seagrass meadow than in bare sand.

Mangrove forests are one of the main imperative and prominent coastal communities of plants in the tropical and subtropical intertidal regions of the world. Because mangroves are halophytes, they require salts to grow and progress. These plants are less competitive in non-saline areas, where marsh plants easily outcompete mangroves. In general, mangroves are further lavish in lower saline areas. Mangroves cannot extend in freshwater environments. Though many mangrove species are able of growing quite well in freshwater, salinity is imperative in reducing competition from other vascular plants. All of this vegetation can grow in areas with low-oxygen soil. It can only reproduce in tropical and subtropical latitudes near the equator because it cannot tolerate cold temperatures. Mangrove forests aid in coastal protection by reducing erosion caused by storm surges, high currents, waves, and tidal amplitude. The mangrove has a complex root system that attracts fish and other organisms looking for food and protection from predators. Several mangroves are predictable by their opaque tangle of prop roots that make the trees emerge to be standing on stilts beyond the water.

Since the Mesozoic era, three groups of flowering plants have colonized the oceans. They are known as "seagrass" because they are the only flowering plants that can survive submerged. They are known as "seagrass" because they are the only flowering plants that can survive submerged. They also absorb a significant amount of nutrients from coastal runoff and sediment, which aids in keeping the water clear. Adapting to a marine environment necessitates significant morphology and structural changes. The geographic distribution and speciation of seagrasses have been influenced by their reliance on seawater. A variety of environmental factors influence whether or not seagrass grows and thrives. Temperature, salinity, waves, currents, and depth all have an impact on the physiological activities of seagrasses. Furthermore, many natural phenomena restrict the photosynthetic activities of plants.

Instead of plants, kelp forests are made up of slightly larger Phaeophyceae. Kelp forests are made up of many different species of kelp. If the surrounding environment is ideal in terms of physical conditions, it can grow up to 45 cm in a single day. The kelp forest is one of the most diverse ecosystems in the ocean. Many fish species, as well as marine birds and mammals such as dugongs and otters, use kelp forests as nurseries for their juveniles. They are used by grey whales to protect themselves from predators and storms. The Kelp forests are currently threatened by warming oceans and out-of-balance ecosystems.

Image 4-4 (A) Seaweed

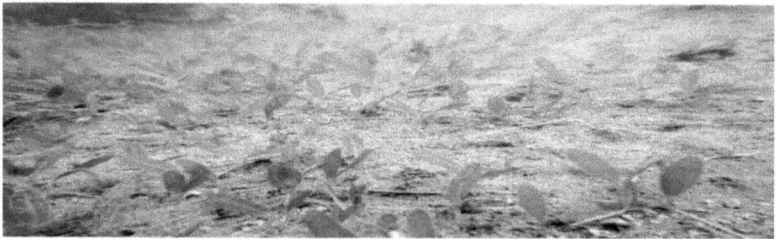

(B) Sea grass (Sources:©DevAdhvan)

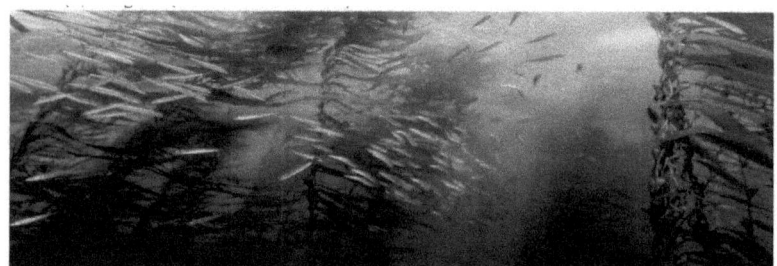

(C) Kelp Forest(Sources: Kelp forest on Cortes Bank, California. National Geographic)

(D) Mangrove

References:

- Adl SM, Simpson AG, Farmer MA, Andersen RA, Anderson OR, Barta JR, Bowser SS, Brugerolle G, Fensome RA, Fredericq S, James TY, Karpov S, Kugrens P, Krug J, Lane CE, Lewis LA, Lodge J, Lynn DH, Mann DG, McCourt RM, Mendoza L, Moestrup O, Mozley-Standridge SE, Nerad TA, Shearer CA, Smirnov AV, Spiegel FW, and Taylor MF 2005."The new higher level classification of eukaryotes with emphasis on the taxonomy of protists". The Journal of Eukaryotic Microbiology. pg. 399–451.
- Andrássy,1992."A short census of free-living nematodes, "Fundamental and Applied Nematologie", pg. 187–188.
- Boca Raton. Carroll, M.L., Denisenko, S.G., Renaud, P.E. & Ambrose, W.G. 2008. Benthic infauna of the seasonally ice-covered western Barents Sea: Patterns and relationships to environmental forcing. Deep Sea Research II pg. 2340-2351.
- Brook, F. J. 1999: The coastal scleractinian coral fauna of the Kermadec Islands, Southwestern Pacific Ocean. Journal of the Royal Society of New Zealand pg. 435–460
- Brusca, R. C. and G. J. Brusca. 2003. The Crustacea. In: Brusca R.C. and G. J. Brusca, Invertebrates. Sunderland, MD: Sinauer Associates; Crustaceans. pg. 511–587
- C. Heip, M. Vincx, and G. Vranken, 1985 "The ecology of marine nematodes," Oceanography and Marine Biology an Annual Review, pg. 399–489.
- Carr, C.M., Hardy, S.M., Brown, T.M., Macdonald, T.A. & He-bert, P.D.N. 2011. A Tri-Oceanic Perspective: DNA Barcoding Reveals Geographic Structure and Cryptic Diversity in Canadian Polychaetes.
- Carroll, M.L. & Carroll, J. 2003. The Arctic Seas. In: K.D. Black & G.B. Shimmield (eds.). Biogeochemistry of Marine Systems, pg. 127-156.

- Carvalho, R., C.L. Wei, G. Rowe and A. Schulze 2013.depth-related patterns in taxonomic and functional diversity of polychaetes in the Gulf of Mexico.
- Chertoprud, E.S., Garlitska, L.A. &Azovsky, A.I. 2010.Large-scale patterns in marine harpacticoid (Crustacea, Copepoda) diversity and distribution. Marine Biodiversity 40: pg. 301-315.
- Clarke, A. 1992. Is there a latitudinal diversity cline in the sea? Trends in Ecology and Evolution: Marineinvertebrates. Pg. 286-287.
- Ereskovsky AV 2010. The comparative embryology of sponges. London, New
- York, Heidelberg: Springer.
- H. M. Platt and R. M. Warwick, 1980 "The significance of free-living nematodes to the littoral ecosystem," in The Shore Environment—2. Ecosystems.
- Hutchings, P. 1998. Biodiversity and functioning of polychaetes in benthic sediments. Biodiversity Conservation pg. 1133–1145.
- Jiang Y, Xu H, Zhu M, Al-Rasheid KAS 2013. Temporal distributions of microplankton populations and relationships to environmental conditions in Jiaozhou Bay, northern China. Journal of Marine Biological Association of UK pg. 13–26.
- Kelly M, Vacelet J 2011. Three new remarkable carnivorous sponges (Porifera, Cladorhizidae) from deep New Zealand and Australian (Macquarie Island) waters. Zootaxa pg. 55–68.
- Kruk C, Mazzeo N, Lacerot G, Reynolds CS 2002. Classification schemes for phytoplankton: a local validation of a functional approach to the analysis of species temporal replacement. Journal of Plankton Research.
- O. Giere, and Meiobenthology 2009: The Microscopic Motile Fauna of Aquatic Sediments, Springer, Berlin, Germany, 2nd edition.

- O'Malley MA, Simpson AG, Roger AJ 2012. "The other eukaryotes in light of evolutionary protistology". Biology & Philosophy. Pg. 299–330.
- Peña Cantero ÁL. 2010. Benthic hydroids (Cnidaria: Hydrozoa) from Peter I Island (Southern Ocean, Ant-arctica). Polar Biology. Pg. 761-773.
- Pocklington, P. and P.G. Wells 1992. Polychaetes key taxa for marine environmental quality monitoring. Marine Pollution Bulletinpg. 593–598.
- Shuchi B, Dodia PP, and Srinivasan M 2017. Intertidal Diversity of Marine Mollusca at RangbaiCoast, Gujarat Journal of Marine Biology and Oceanography.
- Shuchi Bhatt, D. Joshi, R. D. Kamboj 2020. Diversity of Marine Mollusca in Gulf of Kachchh, Gujarat, Environment and Ecology.
- Van Soest RWM, Boury-Esnault N, Hooper JNA, Rützler K, de Voogd NJ, et al. 2011. World Porifera database.
- Voultsiadou E, Vafidis D, Antoniadou C 2008. Sponges of economical interest in the Eastern Mediterranean: an assessment of diversity and population density. Journal of Natural History pg. 529–543.
- Z. X. Chen, S. Y. Chen, and D. W. Dickson, 2004. Eds., Marine Nematode Biodiversity, pg. 436–467, CABI Publishing, Wallingford, UK.

Websites:

- https://www.ck12.org/book/ck-12-biology/section/14.2.
- https://www.seaturtlecamp.com/marine-arthropods/
- https://www.seagrasswatch.org/seagrass/

Chapter 5
Coral Reefs

The 'coral' is derived from the historical Greek word 'korallion', which mentioned the treasured pink coral of the Mediterranean region. The word 'coral' and 'coral reef' both frequently make images of bright green-blue water at bath temperatures, serene shallow lagoons, and seashores of spotless white sand coated with coconut palms. For many people, beliefs of tropical islands evoke images of a special type of coral reef. The term 'Coral' refers to coelenterates predominantly of the order Scleractina. All corals are marine, coelenterates. Most corals are living in colonies of several equal individual components invented of hundreds to thousands of dense units called polyps. The corals develope in warm and temperate seas at a depth ranging from 50 to 10,000ft. Some corals live in the cold water i.e., *Lophelia* which can endure and develop up to 3,300 meters in depth. The corals are classified as cnidarianspecies in two subclasses Hexacorallia and Octocorallia of the class Anthozoa. Hexacoralina consist of the stony corals and these groups have polyps that normally have a six-fold symmetry. Octocoralline includes blue corals and smooth corals.

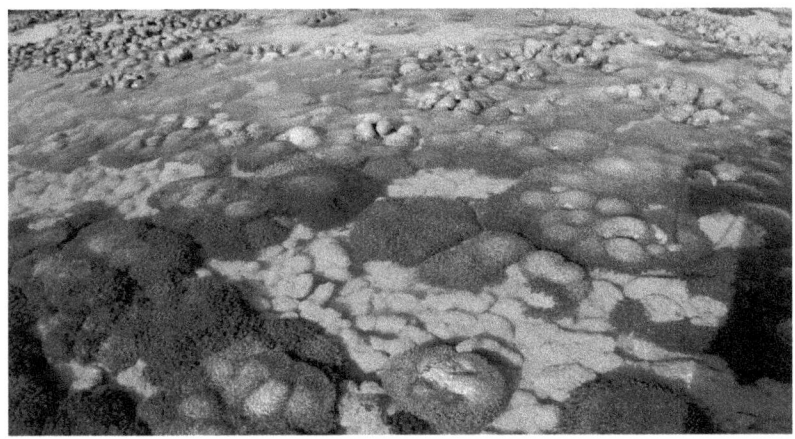

Image 5-1: Incredible scenario of coral reef habitat

5.1 Coral reef Structure and Essential requirements

The coral reef is a compound of various features that are only a fraction of a living coral or algal framework. Though the erstwhile associated features result from this subsist section. The coral reefs are calcareous structure in shallow tropical water that supports various links of marine life forms. An exclusive characteristic of coral reefs which are created by definite of the plants and animals that dwell in them. The mainly vital of these reef-building organisms other than corals, are oysters, polychaete worms, and red algae can form reefs. The most significant of these are algae, which are necessary to reef growth. Several marine biologists believe that coral reefs should be called "algal reefs" or, to be fair to both, "biotic reefs". It is essential to the growth and development of corals. It is important in some places, but they are minor kind compared to coral reefs. But other organisms may be significant; however coral reefs do not build up devoid of corals.

In reef-building corals, the polyps can make skeletons. Billions of these minute skeletons construct an enormous reef. It is found that not all corals are helpful for reefs building. Almost every one of soft corals-order Alcyonacea has a lack of rigid skeleton, though they are copious on coral reefs. These corals are not playing a direct role for reef building them but they are the chief architects. The organisms are accumulations of calcium carbonate or limestone that contribute to modern reef formation. The limestone is absorbed from the water by colonies of polyps and coralline algae. Mainly the original organization of the reef is dead; it's made up of layer upon layer of coral skeletons.

Corals require very explicit environmental conditions that conclude where reefs expand; coral reefs are unusual on soft bottoms. The environmental factor like light, temperature, circulation, sediments, salinity, etc. The corals are growing only in shallow water because of the zooxanthellae and calcareous algae on which they rely on sunlight. Different types of coral and algae have diverse depth limits; some can subsist deeper than others. Overall, the coral reef rarely developed in water deeper than 165 ft. It is also preferred lucid waters since it is cloudy amongst sediment or plankton does not endureluminosity to go in very well.

Hermatypic corals are restricted to temperate water and they can raise and replicate only if the normal water temperature is beyond about 20°C. The majority of reefs are growing within extensively warmer areas, too warm also bad for corals. The upper-temperature boundary varies but is regularly around 30° to 35°C. The exact temperatures stages are desired with the aid of using corals are extraordinary from location to location due to the fact corals from a selected area adapt to the everyday temperatures there. The upper-temperature tolerances of corals are usually now no longer extra ordinarily greater than everyday temperature stages on the location in which they exist. The reef-building coral requires salty water, so it prefers 32 to 42 parts per thousand. Some fine sediment like silt, clay is very destructive to corals. It clouds the water and cuts down light penetration for the zooxanthellae algae. Encrusting coralline algae not only help construct the reef but also assist remain it from washing away. The stony path formed by these algae is harsh sufficient to endure waves that would demolish yet the roughest corals. The algae form a diverse edge on the surface boundary of numerous reefs, mainly in the Pacific area. Encrusting algae carry out however another job that is crucial to reef development.

Coral reef begins to appear when free-swimming coral larvae connect to inundated rocks or solid substrate beside the edges of islands or continents. As the corals lift and increase, reefs are taken on one of three major feature structures; fringing, barrier, or atoll. In addition to being several of the mainly stunning and biologically varied habitats in the ocean, barrier reefs and atolls also are some of the oldest. Corals are begun to inhabit and raise on the fringe of an oceanic island become a fringing reef. It can take as long as 10,000 years for a fringing reef to structure. Above the next 100,000 years, if surrounding circumstances are suitable, the reef will go on to enlarge. As the fringing reef expands, the core island usually begins to collapse and the fringing reef turns into a barrier reef. When the island entirely settles under the water leaving a ring of growing coral with an open lagoon in its center, it is called an atoll. The progression of atoll structure may take as long as 30,000,000 years to occur. With developmental rates of 0.3 to 2 centimeters per year for gigantic corals and up to 10 centimeters per year for branching corals, it can take about 10,000 years for a coral reef to appearas of group larvae. It is depending on their size, barrier reefs and atolls can take from 100,000 to 30,000,000 years to fully structure.

5.2 Coral reef distribution

The coral reefs approximately cover up 111,000 sq.mile. The enormous main stream of great reefs are formed by corals in shallow, sunny waters which are placed inside a tropical zone located between 30° N and 30° S latitude with a favoured temperature variety of about 71° to 84.2° F. diverse species of corals are occurring in a global ocean, from the tropics to the polar regions. Hermatypic corals are increased throughout the tropical and subtropical, Western Atlantic, and Indo-Pacific oceans. Western Atlantic reefs include these areas like Bermuda, Bahamas, Caribbean Islands, Belize, Florida, and the Gulf of Mexico. While in the Indo-Pacific Ocean regions are expanded from the Red Sea and the Persian Gulf throughout the Indian and Pacific oceans to the western coast of Panama.

The Corals are growing preeminent in areas with minute sediment in the water, so great coral reefs structure are not general to locations where there is a huge effort of sediment to the coastal zone by the river. While there is the cold, deep water of coral nearby in the ocean basins, they do not produce huge near-shore reef formations that concern adjacent coasts. The ahermatypic corals are found in the deep-sea and temperate zone as well as in tropical areas. The diversity and distribution of coral reefs are more plentiful and varied in the indo-pacific than in the Atlantic Ocean. The coral reefs are limited to below 30⁰ latitude where annual sea surface temperature. The reef-forming corals generally increase only in warmer water. Perhaps, warm waters can high rates of calcium carbonate deposition needed for hermatypicbe achieved.

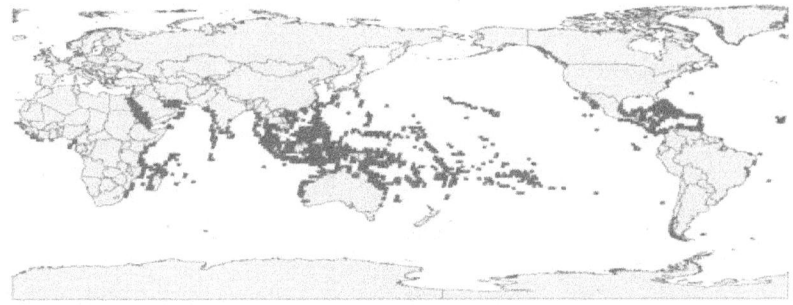

Image 5-2: The Worldwide Coral reef distribution
(Sources: NASA's satellite Landsat 7 spacecraft were collected for the Millennium Coral Reef Mapping Project.)

5.3 Hard corals and Soft corals

The hard coral is also known as Scleractinian corals. The hard corals are completed of severe limestone and emerge very much like rocks. Every polyp secretes a solid exoskeleton made up of limestone and a chalky inner skeleton that stays in place even they die. The coral develops a bit larger because each polyp is so tiny. The hard coral is built for the very subsistence of the reef. In life, they form the living facing of the reef and supply shelter for plentiful other organisms. The breakdown of their skeletal objects after death supplies material for redeployment and consolidation into the reef system. The coral reefs are varied, plentiful, and morphologically variable. The indo-pacific region has about 500-600 species of reef-building hard corals are inhabitant; Staghorn coral, Pillar Coral, Table coral, Brain coral, Blue coral, Elkhorn coral, etc.

The Scleractinian corals are further divided into two groups: the ahermatypic or non-reef building and the hermaphytic or reef-building corals. The scleractinian corals are found in all the world's oceans. It is mainly colonial corals that tend to survive in clear, shallow tropical waters. They are the world's main coral reef builders. Non-reef corals are either colonial or solitary. Some corals are survived in tropical waters but some are dwelling in temperate seas, polar waters or live at immense depths from the photic zone. The hermatypic corals are built reef by hard calcium carbonate material for skeleton and forming the stony framework. Several reef-forming corals have symbiotic-photosynthetic zooxanthellae, but, some zooxanthellae are non-reef-forming corals.

The Soft corals are flexible and frequently resemble plants. These corals do not have rocky skeletons. It has minute, thorny skeletal elements. The soft corals are found around the global ocean, they are mainly found in deep-sea, polar waters, tropics, and subtropics. Many soft corals are straight forwardly collected from the natural habitat for the reef aquarium hobby. Soft corals have less prone to contamination or break during shipping than hard corals. It can grow hastily in captivity and is simply divided into new individual species.

5.4 Types of coral reef

The coral reefs are generally divided into three major groupings: fringing reefs, barrier reefs, and atolls. Some reefs do not well tidily into any exacting category and some scientists have identified further between two categories; bank or platform reef and patch reef. Other some reefs such as Apron reef, Ribbon reef, Table reef, Habili, Micro atolls and Cays, etc. Since Darwin's classification of the three usual reef formations; the fringing reef about a volcanic island becomes a barrier reef and then an atoll.

- *Fringing reefs*

It occurs adjacent to the near shore with modest or no separation from the shore where environmental conditions are suitable. The fringing reefs develop through the rising growth of reef-forming corals on an area of the continental shelf. It often forms a shallow lagoon area between the beach and the main reef area. They bounded islands as well as border continental landmasses and are the most common types of the reef. If a fringing reef grows straight from the shoreline, then the fringing reefs expand to the beach and there is no back reef. In some cases, fringing reefs may rise hundreds of yards from the shoreline and have wide spread back reef areas.

The largest fringing reef is located at Ningaloo Reef, West Australia; it is extended to about 260 km of coastal line. The fringing reefs are found near many tropical areas, it also surrounded several south pacific and Indian Ocean islands. It is commonly found in the Philippines, Indonesia, Western Australia, Caribbean, etc. Two main mechanisms which are make up a fringing reef i.e., Reef Flat and Reef Slope. In the fringing reef, toxic materials are derived from directly dumped which carries many harmful substances.

- *Barrier reefs*

Barrier reefs are widespread linear reef compounds that are similar ashore and are separated from the bordering landmass by a lagoon. It is the chief along with three reefs; it is hundreds of kilometres and many kilometres wide. Barrier reefs are far less ordinary than fringing reefs or atolls, while examples can be found in the tropical Atlantic as well as the Pacific. Some parts of the barrier reef formation are often produced beyond sea level as low-lying coral islands. It develops as wave action deposits coral wreckage broken off from the reef itself. The Great Barrier Reef is the best example, it is covering around 1200-mile off the North-East coast of Australia. The Great Barrier Reef is not a single reef very large compound consisting of many reefs.

(A) Active volcanic eruption   (B) Fringing Reef

(B) Barrier Reef   (D) Atoll

Image 5-3: Above images (A and B) are shown the initial structure of coral reef formation. After Fringing reef formation it turns into Barrier and final Atoll formation.

- *Atoll*

It is regularly spherical or oval with a central lagoon. Atolls comprise ribbons of reefs that may not constantly be spherical but whose broad configuration is a closed shape up to dozens of kilometres crosswise, encircling a lagoon that may be around 160 feet deep or more. There are around 440 atolls are found in the world. Most atolls are found around the Pacific Ocean and the Indian Ocean, somewhat than the Atlantic, because these regions have contained active land margins and subduction zones where volcanic activity is widespread. It also arises as a result of deep-sea volcanic activities and these are much rarer in the Atlantic Ocean. The major atolls are the Caroline Islands, Tokelau, Coral Sea Islands, Marshall Islands, Phoenix Islands, Line Islands, Tuamotu Islands, Kiribati, Northern Cook Islands, Nicaragua Islands, Chagos Archipelago, Lakshadweep, Maldives, etc.

The great Chagos Bank is the largest atoll around the world. It is covering regarding 12,642 square kilometres of area. Most of the fringing reef itself is an underwater characteristic; it is growing from the abyssal floors of the ocean to just beneath high-tide levels. About the edge beside the top of the reef, there are usually low. The basis of atolls has always mesmerized sailors and naturalists, who early appreciated that, though reef-building organisms are, dwell in only the shallowest depths of the sea. The current explanation of atoll formation is incorporate the theory of Darwin, who suggested that atolls are represented the ending stage of a continuing up increase of reef about a sinking extinct volcanic island that had long since disappeared.

- *Patch Reef*

It is isolated patches of corals that are in close nearness to each other but are bodily separated by sand rings. Generally, they are found in shallow lagoons within a larger collective reef or atoll. It is found a depth of ten to twenty feet. The Patch reef is common in the Caribbean islands and the Bahamas, as well as Bermuda. Because patch reefs are in nature a part of larger reef systems that are not traced as individual reefs.

- *Bank Reef*

It is a straight or crescent-shaped outline that is larger than a patch reef. Sometimes it is referred to as platform reefs. They are built rising from the seafloor by non-photosynthetic corals. A solitary species regularly builds these deep-water reefs. Bank reefs are distinctive in that they are found at greater depths than other coral formations; typically dwell at a depth of around 20-60 feet.

- *Microatolls*

These are colonies of corals, commonly *Porties*. It is dead on the upper side but living about its perimeter. It can develop too numerous meters in diameter. Their rising development is inhibited by sea-level through drawn-out revelation at the lowest spring tides. Its dead greater surfaces have been limited by past sea levels. It is valuable because the microatolls are acted as natural recorders of sea-level changes, which is of exacting importance for coral atolls thought to be vulnerable to flood and erosion. Maldives Keeling Islands and Kiribati have found sea-level variations over the past few decades.

- *Sea Mount*

The seamount reef is a large, underwater volcanic mountain that rises around 3,300 feet over the adjacent deep-sea floor. Smaller underwater volcanoes are known as sea knolls while others are known as guyots. It is madeup of dredged material that is microcrystalline or glassy. The oceanic basalt is perhaps formed as submarine lava flows. The Great Table mount is more than 13,120 feet above the nearby landscape.

- *Apron Reef*

The Apron reef is quite similar to the fringing reef, but it is more sloped. It is expanded out and downhill from a point or peninsular shore.

- *Ribbon Reef*

It is a stretched, narrow, perhaps winding reef, usually associated with an atoll.

5.5 Coral reef biodiversity

The coral reefs are some of the world's most dynamic ecosystems. Coral reefs are believed by several scientists to have the key biodiversity of any ecosystem on the earth. The coral reef is one of the most endangered habitats of the earth.

The loss of corals and its associates is impending danger to other biodiversity. It dwells in less than one percent of the floor; it is home to above 25% of all marine life. Every species play its individual function in a coral reef ecosystem. some are herbivores specializing in eating diverse kinds of algae, keeping corals from being smothered by their likely noxious competitors. others, like sharks other predatory fish which maintain populations of lesser fish and other organisms in stability.

The marine formations consist of the enclosed skeletons of living reef-forming organisms and their dead residue. Such reefs were developed by various sessile like oysters, vermetid, serpulid, polychaets, calcifying red algae, and corals. Worldwide approximately 600,000 to more than 9 million species are dwelling in the reef ecosystem. This species diversity is intense in the central Indo-Pacific and declines with long distance from the Indo-Australian archipelago. Concealed underneath the ocean, coral reefs pour with many lives. Fish, corals, worms, lobsters, clams, seahorses, sponges, sea urchins, and sea turtles are only a few of the thousands of animals that rely on reefs for their survival.

5.6 Great Barrier Reef

The Great Barrier Reef is included in the world heritage area; it is one of the main biologically different areas in the world. The Great Barrier Reef is extended 2,100 kilometers from the tip of Cape York along the Queensland coast and covers around 350,000 square kilometres. It contains about 3,000 individual reefs, and 900 islands. This unbelievable ecosystem can even be viewed from space. If you want to get an idea of the size of the great barrier reef, then imagine the country of Italy lying just off the coast of Australia. It has more than 30 species of Mammals (whale, dolphin, and porpoise), 6 species of turtles, 17 species of sea snakes. It is amazingly, 10% of the world's fish species inhabit the Great Barrier Reef. In the great barrier, reef corals are of the precedent could be up to 20 million years old! Some of the coral formations could date back millions of years, but so too could some of the organisms living there.

The nautilus is unique in that, it shows to have remained relatively unchanged over the last 500 million years! The Great Barrier Reef meets to wet tropical area which is the oldest rainforest in the world. It's the only place in the world where two UNESCO World Heritage Sites are meeting. The geological evidence indicated that Australia has moved northward during Cenozoic. The Great Barrier Reef's progression history is complex. The Queensland drifted into tropical waters; it was mainly influenced by reef growth and decline as sea level changed. Terrestrial mark which formed the substrate of the existing great barrier reef was a coastal plain formed the eroded sediments of the great diving range among some larger hills.

Image 5-4: The incredible ecosystem of a great barrier reef

5.7 Coral Triangles

The coral triangle is also known as the 'Amazon of the Ocean'. The Coral Triangle is a marine region that extends six countries from its core; it's an area of an ocean of Indonesia, Malaysia, Papua New Guinea, Philippines, Solomon Islands, and Timor-Leste. The Coral Triangle covers portions of two different biogeographic regions: the Indonesian-Philippines Region and the Far South-western Pacific Region. This magnanimous underwater ecosystem in Southeast Asia is a hotspot for marine biodiversity; it has around 30 percent of the world's coral reefs that cross 6 million square kilometres. It is a large, roughly triangle-shaped marine realm. It is coral reefs that cover about 73,000 square km, about one-third of the world's total.

The Coral Triangle is home to representatives of three-quarters of the world's coral species. It comprises about 600 species of hermaphytic corals, between these, 15 of which are endemic to the area. It is the hometown of 6 out of 7 of the world's marine turtle species, including the hawksbill, leatherback, and loggerhead. Many millions of people also live inside the Coral Triangle and utilize the region's wealth for food and their livelihoods. Like all of our earth's reefs, the Coral Triangle is in danger status. Because of its enormous economic worth, the natural resources found in these waters are exposed to exploitation. Researchers in the reef sciences have recommended that the Coral Triangle was the historical point of origin for countless species of coral on Earth.

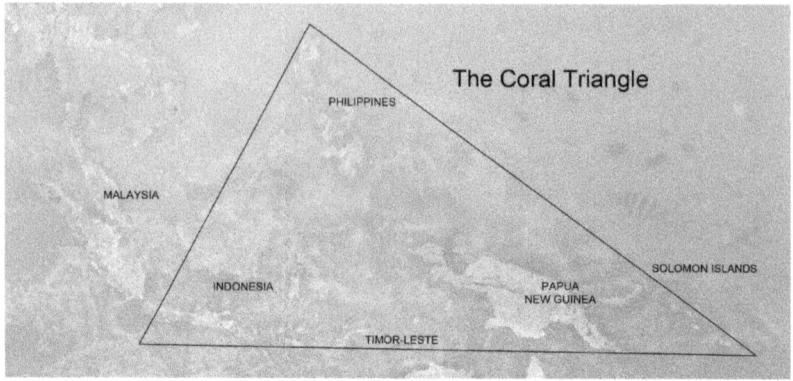

Image 5-5: Coral Triangle with six marginal countries
(Source: Asian Development Bank, 2016)

5.8 Economical value of coral reef

The Coral reefs are remarkably important in terms of the ecosystem services which they transport. It is providing food, living, and economic opportunity to people in more than a hundred countries globally and defends shore from coastal erosion. They are a resource of amusements for national and international tourists alike. Hosting a district of all well-known marine species they also play a significant role in the broader coastal ecosystem. The universal economic value of coral reefs has been approximate at as much as U.S.$ 375 Billion per year. Coral reefs protect coastlines by engrossing wave energy, and various small islands would not exist devoid of their reef to protect them. According to the environmental group World Wide Fund for nature, the economic rate over 25 years of demolished one kilometre of coral reef is somewhere between $ 137,000 and $ 1,200,000. Approximately 6 million tons of fish are taken every year from coral reefs. Well-managed coral reefs have a yearly yield of 15 tons of seafood on common per square kilometre. Southeast Asia's coral reef fisheries only yield about $2.4 billion per annum from seafood industries. Coral reefs are mainly biologically diverse and valuable ecosystems on Earth. Around half a billion people worldwide depend on reef ecosystems for various reasons such as Tourism, Fisheries, Medicine, coastal protection, and other businesses based near coral reef ecosystems. About half of all centralized managed fisheries depend on coral reefs and their related habitats for a part of their life cycles.

The tourism zone contains on-reef tourism which includes snorkelling, SCUBA diving, and boat trips while reef-adjacent tourism is including coastal tourism activities that promote obliquely coral reefs during the provision of calm waters, beaches, and other striking views. On-reef tourism and reef-adjacent both are returns to the tourism and indirect economic multipliers generated crosswise other sectors. Several tourists from about the world tour coral reef destinations every year, the people are attracted by the stunning white sand beaches and balmy, turquoise waters. The economic giving of tourism to coral reefs is approximately $36 billion to the overall economy each year. This revenue supports millions of jobs in numerous restaurants, clubs, hotels, tour operations, transportation companies, etc. The Great Barrier Reef is the best example of an economical as well fishing area. It produces more than 1.5 billion dollars every year for the Australian economy which is from fishing and tourism.

The range of the trade fisheries zone contains the total value of reef-associated detains fisheries in each region. It includes equally the direct returns to the commercial fisheries and the indirect financial multipliers produce obliquely another zone. It does not include the rate of reef-associated small-scale or artisanal fishing. This signified an essential restriction of the study in that it does not detain the economic and social worth of artisanal fishing to other coastal communities. The critical species which depend on the reef ecosystem, like parrotfish play an important role in bringing this much-needed income to coastal communities. Parrotfish is feed on corals, and rupture them down to expel the popular white sand that tourists congregate to. Interestingly, a single big parrotfish can generate 840 pounds of sand each year. The organisms are found on reefs produce chemical composites that have been used in the treatment of many cardiovascular diseases, ulcers, leukemia, lymphoma, and skin cancer, further more, some remain undiscovered. supplementary than half of all new cancer drug research focuses on marine organisms, There is tranquil much to be understood about coral reef organisms, and scientists believe there is still an as set of medical knowledge that remains undiscovered. The conserving coral reefs will likely lead to upcoming medical breakthroughs.

The coral reefs are made up of hard, jagged structures; coral reefs can decrease wave energy by 97% and supply as a natural flood defence beside almost 45,000 miles of coastline universally. Almost 200 million people are depending on coral reefs to defend themselves from storm surges and waves. As climate changes continue to enhance the power of storms and the likelihood of flooding, coastal communities will become still more dependent on their coral reefs for coastal protection. The researches of coral reefs are significant for providing an obvious, scientifically-testable record of climate change over the past million years or so. It includes records of current major storms and human impacts that are evidenced by the changes in coral growth patterns.

5.9 Threats of coral reef

The coral reefs are important for many reasons apart from obviously containing the diversity of life. Many scientists estimate that there may be millions of undiscovered species of marine organisms living in and about reefs. The vigorous reef ecosystem supports commercial and subsistence fisheries as well as employment and businesses through tourism and amusement. The coral reefs are buffer shorelines against 97 percent of the energy from sea waves, storms, and floods. The reefs are helping to avoid loss of life, property damage, and erosion. When coral reefs are broken or smashed; the lack of this natural barrier can raise the damage to coastal communities from common wave action and violent storms. Despite their immense economic and leisure value; coral reefs are rigorously threatened by other factors such as pollution, disease, habitat destruction, etc. Once the reefs are damaged, it is less able to sustain the several organisms that inhabit them and the communities near them. The coral reefs are a source of nitrogen and other vital nutrients for marine food chains. It supports carbon and nitrogen-fixing and helps with nutrient recovery.

Coral reefs are being degraded by an increase of stresses arising from different human activities. Many natural, as well as manmade factors are responsible for coral reef damage. Overfishing, pollution, and coastal progress are top-listed chronic stressors. The chronic stresses are devastating the flexibility, of reef communities. Some coral reefs are covered up with sand, rock, and concrete to make low-priced terrain and stimulate economic progress. The long-term transformation in the oceans and atmosphere, and acute stresses from extremely changeable seasons, predators, severe storms, earthquakes, and volcanic eruptions also influence coral reefs.

World wide disasters such as storms and earthquakes occur naturally and regularly. These natural disasters are further severe if reef communities are already damaged by other impacts and healing is withdrawn by seaweed overgrowth due to the lack of grazing organisms, removed by fishing. The natural predators include predators such as parrotfish, barnacles, crabs, and crown-of-thorns starfish. Hurricanes or prolonged cold and rainy weather can harm coral reefs. The El Niño weather, which can result in inferior sea level, altered salinity due to too much rainfall, and elevated sea-surface temperatures can also harm coral.

The abnormal climatic manners can affect stress on coral reefs. The vivid effects of El Niño have increased concern above the effect of climate change on a coral reef. The zooxanthellae algae start to expel; in their tissues causing the coral to turn wholly white. In this circumstance the coral is bleached but it is not lifeless. Corals can live a bleaching event, but they are under more stress and it starts a matter of mortality.

Human activities are chief threats to coral reefs. Pollution, overfishing, critical fishing are using dynamite or cyanide, one of the most important threats to reefs is pollution. Terrestrial runoff and pollutant discharge can result from coastal development, farming, deforestation actions, and sewage action plant operations. It may have sediments, nutrients, insecticides, oil, and debris. When some dangerous pollutants penetrate water it can increase nutrient levels, promoting the swift development of algae and other organisms that can smother corals. The coral reefs are affected by leaking fuels, anti-fouling paints and coatings, and some chemicals that go into the water. Oil spills do not constantly emerge to concern corals directly because the oil generally stays by the surface water.

References:

- ADB 2014. Economics of fisheries and aquaculture in the Coral Triangle. '
- Adhavan, D., R. Chandran, S. Tikadar& K. Sivakumar 2017. Trematode infestation in coral colonies at Poshitra Reef, Gulf of Kachchh Marine National Park, Gujarat, India. Journal of Threatened Taxa 9(6): pg. 10345–10346.
- Bouchet P 2006. The magnitude of marine biodiversity. In: Duarte CM, editor. The exploration of marine biodiversity:

- scientific and technological challenges. Bilbao, Spain: Fundación BBVA. pg. 31–64.
- Bruno, J.F., Selig, E.R. 2007. Regional Decline of Coral Cover in the Indo-Pacific: Timing, Extent, and Subregional Comparisons. PLoS ONE, 2 (8) e711.
- Burke L., Selig, E. & Spalding, M. 2002. *Reefs at risk in Southeast Asia.* World Resources Institute, Washington DC, USA.
- Carpenter KE, Abrar M, AebyG, Aronson RB, Banks S, et al. 2008. One-third of reef-building corals face elevated extinction risk from climate change and local impacts. Science 321: 560–563.
- Cesar, H. 1996. Economic Analysis of Indonesian Coral Reefs. The World Bank.
- Deloitte. 2017. At What Price? The Economic, Social, and Icon Value of the Great Barrier Reef.
- Hughes TP, Bellwood DR, Connolly SR 2002. Biodiversity hotspots, centers of endemicity, and the conservation of coral reefs. Ecology Letters 5: pg. 775–784.
- Idjadi JA, Edmunds PJ 2006. Scleractinian corals as facilitators for other invertebrates on a Caribbean reef. Marine Ecology Progress Series 319: pg. 117–127.
- Jones GP, McCormick MI, Srinivasan M, Eagle JV 2004. Coral decline threatens fish biodiversity in marine reserves. ProcNational Academy of Science USA 101: pg. 8251–8253.
- Joshi, D., Banerji, U., and Mankodi, P.C., 2015, Delayed recovery in Porites spp. following mass coral bleaching: a case study from the Gulf of Kachchh region, Gujarat, India. Journal of Global Bioscience, pg. 2326-2331.
- LaetitiaPlaisance, M. Julian Caley, Russell E. Brainard, Nancy Knowlton 2011. The Diversity of Coral Reefs: What Are We Missing? PLOS ONE
- MohitArora, Nandini Ray Chaudhury, AshwinGujrati, R.D. Kamboj, Devanshi Joshi,

- Harshad Patel and Rakesh Patel 2019 Coral bleaching due to increased sea surface temperature in Gulf of Kachchh Region, India, during June 2016, Indian Journal of Geo-Marine Sciences Vol. 48 (03), pg. 327-33.
- Mondal T, Raghunathan C & Ramakrishna, Addition of thirteen Scleractinians as new to Indian water from Rutland Island, Andamans, 2017. Asian Journal of Experimental Biological Science, 2 (3) pg. 383-390.
- Pastorok, R., Bilyard, G. 1985. Effects of Sewage Pollution on Coral-Reef Communities. Marine Ecology Progress Series, 21, pg. 175-189.
- Rinkevich, B. 2005. What do we know about Eilat (Red Sea) reef degradation? A critical examination of the published literature. *Journal of Experimental Biology and Ecology,* **327**, pg. 183-200
- Small A, Adey A, Spoon D 1998 Are current estimates of coral reef biodiversity too low? The view through the window of a microcosm. Atoll Res Bull 458: pg. 1–20.
- Spalding, M.D., Brumbaugh, R.D., Landis, E. 2016. Atlas of Ocean Wealth. The Nature Conservancy. Arlington, VA.
- Teh, L.S.L., Teh, L.C.L., Sumaila, U.R. 2013.A Global Estimate of the Number of Coral Reef Fishers. PLoS ONE, 8 (6): e65397.
- Toronto and Region Conservation Authority 2010. Low impact development stormwater management planning and design guide.
- UN Environment 2017. Coral Bleaching Futures.
- Venkataraman K &Rajan R, 2013. Status of coral reef in Palk bay *Records of the Zoologicaley of India,* 113 (2) pg. 1-11.
- Verhoeven, J.T.A., Beltman, B., Bobbink, R., Whigham, D.F. 2006. Wetlands and Natural Resource Management. Ecological Studies, Vol. 190. Germany, Springer.

Websites:

- https://www.queensland.com/au/en/places-to-see/experiences/great-barrier-reef/facts-about-the-iconic-great-barrier-reef
- https://www.noaa.gov/education/resource-collections/marine-life/coral-reef-ecosystems
- https://www.britannica.com/science/coral-reef
- https://ocean.si.edu/ocean-life/invertebrates/corals-and-coral-reefs
- https://traveltips.usatoday.com/factors-affecting-coral-reefs-63294.html
- https://www.coraltrianglecenter.org/
- https://www.adb.org/multimedia/coral-triangle/